Mind On Statistics

FOURTH EDITION

Jessica M. Utts
University of California, Irvine

Robert F. Heckard
Pennsylvania State University

Prepared by

Jessica M. Utts
University of California, Irvine

Robert F. Heckard
Pennsylvania State University

BROOKS/COLE
CENGAGE Learning™

Australia • Brazil • Japan • Korea • Mexico • Singapore • Spain • United Kingdom • United States

For product information and technology assistance, contact us at **Cengage Learning Customer & Sales Support, 1-800-354-9706**

For permission to use material from this text or product, submit all requests online at **www.cengage.com/permissions** Further permissions questions can be emailed to **permissionrequest@cengage.com**

ISBN-13: 978-0-538-73604-6
ISBN-10: 0-538-73604-6

Brooks/Cole
20 Channel Center Street
Boston, MA 02210
USA

Cengage Learning is a leading provider of customized learning solutions with office locations around the globe, including Singapore, the United Kingdom, Australia, Mexico, Brazil, and Japan. Locate your local office at: **www.cengage.com/global**

Cengage Learning products are represented in Canada by Nelson Education, Ltd.

To learn more about Brooks/Cole, visit **www.cengage.com/brookscole**

Purchase any of our products at your local college store or at our preferred online store **www.cengagebrain.com**

Printed in the United States of America
1 2 3 4 5 6 7 15 14 13 12 11

CONTENTS

INTRODUCTION

What is This Manual?

This *Student Solutions Manual* is intended to supplement the brief answers provided for odd-numbered exercises in the back of the textbook. It includes fully worked solutions to those exercises.

Tips for Doing Exercises Assigned for Homework

1. You instructor will most likely assign exercises that reinforce ideas and methods he or she thinks are important. Therefore, your first strategy for tackling problems assigned for homework is to review your class notes. You will probably find that the instructor either worked similar examples, or at least provided you with the necessary tools for working out the assigned exercises.

2. Often there will be very similar exercises near each other, especially for the "Skillbuilder" exercises. The odd-numbered exercises with numbers in boldface have answers in the back of the book and are worked out in this *Manual*. Therefore, you might see if there is a boldfaced exercise that is similar to the one you are trying to do, and look at how it is done.

3. The majority of the exercises in the book are listed under specific sections. Therefore, if the exercises assigned are listed as applying to a certain section of the book, look in that section for information on how to solve them. If you are assigned exercises listed under "Chapter Exercises" you can see if there is a similar exercise listed under one of the sections. If so, then you know the relevant material will most likely be found in that section.

4. Whether you can identify the relevant section of the chapter for a particular exercise or not, the next strategy is to look for appropriate "Key Terms." These are listed just prior to the beginning of the exercises in each chapter, and are divided by sections. If you can identify a key term appropriate for a particular exercise, then you can read the material surrounding that term in the book and will most likely find the necessary information to solve the exercise. For instance, suppose you are asked to do Exercise 8.113 in the Chapter Exercises for Chapter 8. The exercise begins "The vehicle speeds at a particular interstate location are described by a normal curve." Looking over the Chapter 8 Key Terms, you will see that the term "normal curve" appears under Section 8.6. If you look in that Section, you will not only find the relevant material, you will also find examples similar to what you are asked to do in Exercise 8.113.

Exercises Not Assigned for Homework

Working exercises is an excellent method for studying statistics. If you are using exercises for that purpose you will be better served by *not* isolating the exercise as applying to a specific section of the book until you understand more about the exercise. One of the most difficult problems students have in solving problems is identifying *which* technique is appropriate. The mechanics are often easy once the appropriate topic has been identified.

Therefore, when you are studying a particular chapter, start by trying to solve the Chapter Exercises that have solutions provided in the back of the book or this *Manual*. Try to get as far as possible before looking at the solution. If you are really stuck, then try identifying pairs of exercises that are similar. Use the solution provided to figure out how to solve the first one, then attempt to solve the second one on your own. Another strategy is to first do the exercises in each section that have solutions provided. Then, attempt to do the Chapter Exercises without looking at the solutions until you are finished.

CHAPTER 1
ODD-NUMBERED SOLUTIONS

1.1   **a.** The fastest speed was 150 miles per hour.

        b. The slowest speed driven by a male was 55 miles per hour.

        c. 1/4 of the females reported having driven at 95 miles per hour or faster. Notice that 95 mph is the *upper quartile* for females. By definition, about 1/4 of the values in a data set are greater than the upper quartile.

        d. 1/2 of the females reported having driven 89 mph or faster. Notice that 89 mph is the *median* value.

        e. 1/2 of 102 = 51 females have driven 89 mph or faster.

        Note: For parts (d) and (e) the answer would have to be adjusted if there were any females who reported 89 as their value, but from the data on page 2 we can see that there were not. Because there were no "ties" with the median, we know that exactly half of the values fall above it and half fall below it.

1.3   **a.** The observed rate of cervical cancer in Vietnamese American women is 86 per 200,000. This could also be expressed as 43 per 100,000 or 4.3 per 10,000, and so on. In decimal form, it is .00043.

        b. The risk of developing cervical cancer for Vietnamese American women in the next year is 86/200000 = .00043.

        c. The rate of 86 per 200,000 is based on past data and tells us the number of Vietnamese American who developed cervical cancer out of a population of 200,000. The risk utilizes the rate from the past to tell us the future likelihood of cervical cancer in other Vietnamese American women.

1.5   **a.** All teens in the U.S. at the time the poll was taken.

        b. All teens in the U.S. who had dated at the time the poll was taken.

1.7   **a.** All adults in the U.S. at the time the poll was taken.

        b. $\dfrac{1}{\sqrt{1048}} = .031$ or 3.1%

        c. 34% ± 3.1%, or 30.9% to 37.1%.

1.9   Solve for n in the equation $\dfrac{1}{\sqrt{n}} = .05 = \dfrac{1}{20}$. Answer is $n = 400$ teenagers.

1.11   **a.** This is an example of a self-selected or volunteer sample. Magazine readers voluntarily responded to the survey, and were not randomly selected.

        b. These results may not represent the opinions of all readers of the magazine. The people who respond probably do so because they feel stronger about the issues (for example, violence on television or physical discipline) than the readers who do not respond. So, they may be likely to have a generally different point of view than those who do not respond.

1.13   **a.** Randomized experiment (because students were randomly assigned to the two methods).

        b. Observational study (because people cannot be randomly assigned to smoke or not).

        c. Observational study (because people cannot be randomly assigned to be a CEO or not).

1.15   Answers will vary, but one possibility is general level of activity. It is likely to differ for elderly people who attend church regularly and those who don't, and it is also likely to affect blood pressure. So it might partially explain the results of this study.

1.17   You would need to know how large the difference in weight loss was for the two groups. If the difference in weight loss is very small (but not 0) it could be statistically significant, but not have much practical importance.

1.19   You would want to know how many different relationships were examined. If this result was the only one that was statistically significant out of many examined, it could easily be a false positive.

1.21   The placebo group estimates the baseline rate of heart attacks for men not taking aspirin. So, the estimated baseline rate of heart attacks is 189/11,034, which is about 17 heart attacks per 1,000 men or 17/1000. (See Table 1.1 for the data.)

1.23   **a.**

	Minutes of exercise per week	
Median	180	
Quartiles	37	330
Extremes	0	600

To determine the summary, first write the responses in order from smallest to largest.
The ordered list of data is:
    0, 0, 14, 60, 90, 120, 180, 240, 300, 300, 360, 480, 600
Minimum = 0 min.
Maximum = 600 min.
Median = 180 min.   (middle value in the ordered list)
Lower quartile = 42 min. It is the median of the values smaller than the median.
    These are 0, 0, 14, 60, 90, 120.
    Median of these six values is (14+70)/2 = 42.
Upper quartile = 330 min. It is the median of the values larger than the median.
    Values larger than the median are 240, 300, 300, 360, 480, 600.
    Median of these values is (300+360)/2 = 330.
b. Reported exercises hours per week for the men in the sample ranged from a low of 0 to a high of 600 minutes per week. The median response was 180 min (3 hours). About 1/2 of the men (the middle half) reported exercising between 37 and 330 minutes (5 and a half hours) per week. About 1/4 said they exercised less than 37 minutes per week while 1/4 said they exercised more than 330 minutes per week.

1.25   **a.** This is an observational study because vegetarians and non-vegetarians are compared and these groups occur naturally. People were not assigned to treatment groups.
b. Since this is an observational study and not a randomized experiment, we cannot conclude that a vegetarian diet causes lower death rates from heart attacks and cancer. Other variables not accounted for may be causing this reduction.
c. This answer will differ for each student. One potential confounding variable is amount of exercise. This is a confounding variable because it may be that vegetarians also exercise more on average and this led to lower death rates from heart attacks and cancer.

1.27   The base rate or baseline risk is missing from the report. You need to know the base rate of cancer of the rectum for men to decide if the increased risk from drinking beer is large or small.

1.29   For Caution 1: Because the difference given in the previous exercise is a large difference (46% with nicotine patch and 20% with placebo), it has practical importance as well as statistical significance. For Caution 2: Because the result is based on a randomized experiment, it is not possible that whether someone quit or not influenced the type of patch they were assigned.

1.31   Neither caution applies. The magnitude of the difference is given in Case Study 1.6, and considering the number of men between the ages of 40 and 84 in the United States population, the given difference has practical importance. Because men were randomly assigned to take aspirin (or not) we can conclude that the correct direction of the cause and effect is that taking aspirin caused the reduction in heart attacks

1.33   **a.** The margin of error is about $\frac{1}{\sqrt{n}} = \frac{1}{\sqrt{1525}} = .026$.

b $.139 \pm .026$, which is .113 to .165.
This is *sample proportion ± margin of error.*

1.35   **a.** This is a self-selected (or volunteer) sample.

b. Probably higher, because people who would say they have seen a ghost would be more likely to call the late-night radio talk show than others. They might even be more likely to be listening to such a show.

1.37 The term "data snooping" refers to looking at the data in a variety of ways until something interesting to report emerges.

1.39 In some situations it is not practical or even possible to conduct a randomized experiment. For example, a researcher may wish to study whether occupational exposure to asbestos affects the risk of lung disease. It would not be possible, or ethical, to assign people to occupations that involve differing amounts of exposure to asbestos.

1.41 The answer will differ for students, but here is an example. Randomly assign volunteers to either eat lots of chocolate or not eat any chocolate for a period of time, and give them a questionnaire about depression at the beginning and the end of the time period. Then compare the change in depression scores for the two groups.

1.43 *USA Today* made the mistake of making a cause-and-effect conclusion about the relationship between prayer and blood pressure. This conclusion is not justified because the data were from an observational study. Specifically, they neglected to consider possible confounding variables like lifestyle choices, social networks, and health of the people between the two groups. As a result, people may be led to believe that if they pray more often they will have lower blood pressure. That conclusion is not justified based on this observational study.

CHAPTER 2
ODD-NUMBERED SOLUTIONS

2.1   **a.** 4
   b. A state in the United States.
   c. $n = 50$.

2.3   **a.** Whole population.
   b. Sample

2.5   **a.** Population parameter.
   b. Sample statistic.
   c. Sample statistic.

2.7   **a.** Sex and self-reported fastest ever driven speed.
   b. Students in a statistics class.
   c. If students represent a larger group of individuals, it is sample data. If interest is only in this group of students, it is population data.

2.9   This is a population summary if we restrict our interest only to the fiscal year 1998. (If we were to use this value to represent errors in other years, it could be considered to be a sample summary.)

2.11   **a.** Categorical.
   b. Quantitative.
   c. Quantitative.
   d. Categorical.

2.13   **a.** Not ordinal. It's categorical but the categories are not ordered.
   b. Ordinal. Grades are ordered categories.
   c. Not ordinal. It's quantitative.

2.15   **a.** Explanatory variable is score on the final exam; response variable is final course grade.
   b. Explanatory variable is sex; response variable is opinion about the death penalty.

2.17   **a.** Not continuous. A student could not miss 4.631 classes for example.
   b. Continuous. With an accurate enough measuring instrument, any measurement is possible.
   c. Continuous. With an accurate enough time piece, any length of time is possible.

2.19   **a.** Whether a person supports the smoking ban or not is a categorical variable.
   b. Gains on verbal and math SATs are quantitative variables.

2.21   **a.** Sex and pulse rate.
   b. Sex is categorical, pulse rate is quantitative.
   c. Is there a difference between the mean pulse rates of men and women? The sample mean pulse rate for each sex would be useful.

2.23   This will differ for each student. One example where numerical summaries would make sense for an ordinal variable is the response to the question "What grade do you expect in this class? 1=A, 2=B, 3=C, 4=D, 5=F." The mean numerical response is an expected class GPA.

2.25   **a.** A unit is a person. Dominant hand is a categorical variable and IQ is a quantitative variable. Explanatory variable is dominant hand and response variable is IQ .
   b. A unit is a married couple. Eventual divorce status and pet ownership are both categorical variables. Explanatory variable is pet ownership and response variable is eventual divorce status.

5

2.27 **a.** 1427/2530 = .564, which is 56.4%.

b. 1 – (1427/2530) = .436, which is 43.6%.

c. Never: 105/2530 = .042 (4.2%); Rarely: 248/2530 = .098 (9.8%); Sometimes: 286/2530 = .113 (11.3%). Most times: 464/2530 = .183 (18.3%); Always: 1427/2530 = .564 (56.4%)

d.

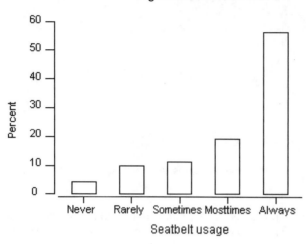

Figure for Exercise 2.27d

2.29 **a.**

	Preferred use of cell phone		
	To talk	**To text**	**Total**
Women	22 (20.8%)	84 (79.2%)	106 (100%)
Men	34 (41.0%)	49 (59.0%)	83 (100%)

b. Women: 20.8% to talk, 79.2% to text

c. Men: 41.0% to talk, 59.0% to text

d. Women were more likely to say "to text" than men whereas men were more likely to say "to talk."

2.31 **a.** Explanatory variable is whether a person smoked or not. Response variable is whether they developed Alzheimer's or not.

b. Explanatory variable is political party. Response variable is whether a person voted or not.

c. Explanatory variable is income level. Response variable is whether a person has been subjected to a tax audit or not.

2.33 **a.** The explanatory variable is sex and the response variable is how they feel about their weight.

b.

Feelings About Weight				
Sex	**Overweight**	**About right**	**Underweight**	**Total**
Female	38 (26.6%)	99 (69.2%)	6 (4.2%)	143
Male	18 (23.1%)	35 (44.9%)	25 (32.1%)	78

c. Feeling overweight: 38/143 = .266, or 26.6%; right weight: 99/149 = .692, or 69.2%; underweight: 6/149 = .042, or 4.2%.

d. Feeling overweight: 18/78 = .231, or 23.1%; right weight: 35/78 = .449, or 44.9%; underweight: 25/78 = .321 or 32.1%.

e. Males are more likely than females to feel that they are underweight; females are more likely than males to say that their weight is about right..

6

2.35   **a.**

	Picked S	Picked Q	Total
S listed first	61	31	92
Q listed first	45	53	98
Total	106	84	190

b. Picked S $= (61/92) \times 100\% = 66.3\%$; Picked Q $= (31/92) \times 100\% = 33.7\%$;

c. Picked S $= (45/98) \times 100\% = 45.9\%$; Picked Q $= (53/98) \times 100\% = 54.1\%$;

d.

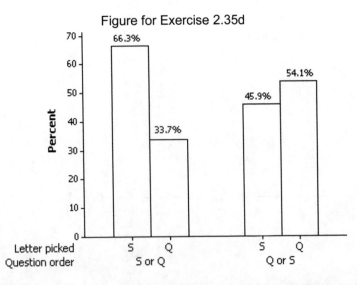

Figure for Exercise 2.35d

e. Parts (b) and (c) show that the percentage picking S was higher when S was listed first than when Q was listed first. It looks like the letter picked was influenced by the letter listed first.

2.37   **a.** The fastest speed was 150 miles per hour.
b. The slowest speed driven by a male was 55 miles per hour.
c. 1/4 of the females reported having driven at 95 miles per hour or faster. Notice that 95 mph is the *upper quartile* for females. By definition, about 1/4 of the values in a data set are greater than the upper quartile.
d. 1/2 of the females reported having driven 89 mph or faster. Notice that 89 mph is the *median* value.
e. 1/2 of 102 = 51 females have driven 89 mph or faster.

2.39   **a.** The center for the females is at a greater percentage than it is for the males. For females the center looks to be somewhere around 27%. For males, the center looks to be a bit less than 18%.
b. The data are more spread for the females.
c. The greatest two female percentages are set apart from the bulk of the data. The values are about 65% and 72%.

2.41   **a.** The median value, 65 inches, describes the location.
b. The interval described by the extremes, 59 to 71 inches, describes spread. We might also describe spread using the interval 63.5 to 67.5, the spread of the middle 50% of the data.

2.43   **a.** The dataset looks approximately symmetric and bell-shaped.
b. There are no noticeable outliers.
c. The most frequently reported value for sleep was 7 hours.
d. Roughly 14 or so students said they slept 8 hours the previous night.

2.45   **a.** The dataset looks roughly symmetric.
b. The highest temperature is 92°F.
c. The lowest temperature is 64°F.
d. 5/20 = .25, which is 25%.

2.47   **a.** The figure below uses separate stems for last digits 0-4 and 5-9. That's not imperative, although doing so give more detail of the shape of the data.

Figure for Exercise 2.47a

```
|5|  6
|6|  2
|6|  8
|7|
|7|  55
|8|  02344
|8|  77
|9|  13
|9|  5
```

b.

Figure for Exercise 2.47b

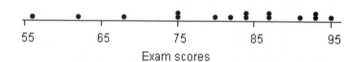

Exam scores

2.49   **a.** In the figure shown here, two stems have been used for each possible "tens" place in the number (values under 10 are an exception because the lowest value is about 6 inches). We rounded the 1995 total of 24.5 inches up to 25.

Figure for Exercise 2.49a

```
|0|  689
|1|  11111122233444
|1|  566667777779
|2|  011
|2|  5555778889
|3|  0011
|3|  7
```

b.

Figure for Exercise 2.49b

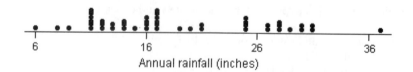

Annual rainfall (inches)

c. The data are skewed (but only slightly) to the right.

2.51   Yes, a stem-and-leaf plot provides sufficient information to determine whether a dataset contains an outlier. Because all individual values are shown, it is possible to see whether are any values are inconsistent with the bulk of the data.

8

2.53   Skewed to the left. Along a number line, values stretch more toward the left (small values).

2.55   **a.** Histogram is better than a boxplot for evaluating shape.
b. A boxplot is useful for identifying outliers, evaluating spread, and for comparing groups.

2.57   Generally, females tended to have higher tip percentages. The median is clearly greater for females. The data for the females also shows greater spread than the data for the males.

2.59   **a.** Mean = 74.33; median = 74.
b. Mean = 25; median = 7.
c. Mean = 27.5; median = 30.

2.61   **a.** $225 - 123 = 102$ lbs
b. $190 - 155 = 45$ lbs
c. 50%

2.63   **a.** Min = 109, $Q_1 = 180.75$, Median = 186, $Q_3 = 199.0$, Max = 214.0
The data in order are 109.0, 178.5, 183.0, 185.0, 186.0, 188.5, 194.5, 203.5, 214.0.
Median is the middle value (= 186). Lower quartile is median of 109.0, 178.5, 183.0, 185, so equals (178.5 + 183).2 = 180.75. Upper quartile is median of 188.5, 194.5, 203.5, 214, so equals (194.5 + 203.5)/2 = 199.0
b. 109.0 is an outlier. It's below the lower outlier boundary = $Q_1 - 1.5\, IQR = 180.75 - 1.5\,(199 - 180.75) = 153.375$.
c. The member who weighed 109.

2.65   The boxplots below show that, on average, the fastest speeds ever driven by males tend to be higher than the fastest speeds ever driven by females. It is also seen, if outliers are ignored, that the spread is greater for males than it is for females. A horizontal axis has been used for fastest speeds here, but a vertical axis would be equally appropriate.

Figure for Exercise 2.65

2.67   **a.**

	Minutes of exercise per week		
Median		180	
Quartiles	37		330
Extremes	0		600

To determine the summary, first write the responses in order from smallest to largest.

9

The ordered list of data is:

0, 0, 14, 60, 90, 120, 180, 240, 300, 300, 360, 480, 600

Minimum = 0 min.

Maximum = 600 min.

Median = 180 min.   (middle value in the ordered list)

Lower quartile = 42 min. It is the median of the values smaller than the median.
  These are 0, 0, 14, 60, 90, 120.
  Median of these six values is (14+60)/2 = 37.

Upper quartile = 330 min. It is the median of the values larger than the median.
  Values larger than the median are 240, 300, 300, 360, 480, 600.
  Median of these values is (300+360)/2=330.

b. All men in the sample reported exercising between 0 and 600 minutes per week. The median response was 180 min. About 1/2 of the men reported exercising between 37 and 330 minutes per week. About 1/4 said they exercised less than 37 minutes per week while 1/4 said they exercised more than 330 minutes per week.

c.

Figure for Exercise 2.67c

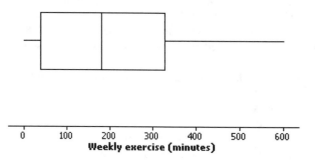

2.69   The median annual rainfall in Davis, CA is 16.72 inches and the mean is 18.62 inches. The data values vary from 6.14 inches (in 1965) to 37.42 inches (in 1982). Although not an extreme outlier, the 1982 value is separated somewhat from the other values. The 1982 value of 37.42 inches is about 6 inches more than the next highest total, which occurred in 1981. It is interesting to note that the two highest values were in consecutive years.

2.71   There are n =47 values so the median is 24th value in the ordered data (23 will be smaller and 23 will be larger). The lower quartile is the median of the smallest 23 values and the upper quartile is the median of the largest 23 values. These will be, respectively, the 12th lowest and the 12th highest values in the data set. The five-number summary is:

	Rainfall (inches)		
Median		16.72	
Quartiles	12.05		25.37
Extremes	6.14		37.42

The five-number summary above shows that the median annual rainfall for Davis, California is 16.72 inches. The middle ½ of the values are between 12.05 and 25.37 inches. The minimum is 6.14 inches and the maximum is 37.42 inches.

2.73   Although the raw data are not available, we can determine that the median is somewhere between 21 and 25 words. There are n=600 sentences, so the median occurs between the 300th and 301st observations in the ordered data. The total number of sentences with 20 or fewer words is the sum of the frequencies for categories up to and including the 16 to 20 words category. This is (3+27+71+113) =224 sentences. There are 107 sentences in the 21 to 25 words category, so there are 224+107 = 331 with 25 or fewer words, The 300th and 301st values must be somewhere between 21 and 25 words.
Each quartile is the average of the 150th and 151st value from the appropriate side of the ordered data. Using the same reasoning that we did for the median, the first quartile must be somewhere between 16 and 20

10

words, and the third quartile is somewhere between 31 and 35 words. As a result, the *IQR* might be as low as 31−20=11 words or as high as 35−16= 19 words.

The exact values of the minimum and maximum aren't given, but the minimum may be about 3 or so and the maximum might be about 57, the range is likely to be somewhere around 57−3=54.

2.75   **a.** The median is the average of the middle two values. There were *n*=50 ages, so the lower quartile is the median of the 25 lowest ages and the upper quartile is the median of the 25 highest ages.

	CEO ages (years)		
Median		57.5	
Quartiles	52		62
Extremes	42		78

The five-number summary shows that the median age of the 50 CEOs is 57.5 years. The middle ½ of the CEOs have ages between 52 and 62 years. The youngest CEO is 42 years old. The oldest CEO is 78 years old.

2.77   The mean of the CEO ages is 57.84 years. The median is 57.5 years. The mean and the median are similar. This is expected because the data are more or less symmetric in shape.

2.79   This will differ for each student. One example is that a person 80 years old would be an outlier at a traditional college, but would not be an outlier at a retirement home.

2.81   The rainfall total for 1982 is high (37.42 inches) but not high enough to be classified as an outlier. There are no potential outliers at the lower end.

2.83   Most likely a mistake was made when the data were entered. If possible, the instructor should correct the value (by looking again at the student's survey form). If the correct height is not available, the value 17 should be deleted from the dataset.

2.85   **a.** Mean ± St. Dev is 7 ± 1.7, or 5.3 to 8.7.
        b. Mean ± 2 St. Dev is 7 ± (2)(1.7), or 3.6 to 10.4.
        c. Mean ± 3 St. Dev is 7 ± (3)(1.7), or 1.9 to 12.1.

2.87   **a.** Mean = 25, s = 4.24. Calculation of s is is is $s = \sqrt{\dfrac{(22-25)^2 + (27-25)^2 + (30-25)^2 + (21-25)^2}{4-1}}$

        b. Mean = 30, s = 9.13. Calculation of s is is is $s = \sqrt{\dfrac{(25-30)^2 + (35-30)^2 + (40-30)^2 + (20-30)^2}{4-1}}$

2.89

Figure for Exercise 2.89

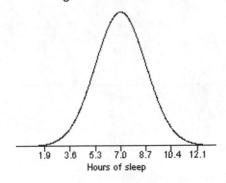

Hours of sleep

2.91   **a.** $z = (300-350)/100 = -0.5$.
        b. $z = (460-350)/100 = 1.1$.
        c. $z = (650-350)/100 = 3.0$.
        d. $z = (210-350)/100 = -1.4..$

2.93   The only possible set of numbers is {50, 50, 50, 50, 50, 50, 50} because a standard deviation of 0 means there is no variability.

2.95   **a.** $98-41 = 57$.
        b. Standard deviation $\approx$ Range/6 $= 57/6 = 9.5$.

2.97   The Empirical Rule says that 68% of values fall within 1 standard deviation of the mean, 95% fall within 2 standard deviations of the mean, and 99.7% fall within 3 standard deviations of the mean. Of the 103 hand-span measurements for women, 74 or 72% are within 1 standard deviation of the mean (18.2 to 21.8 cm). 100 of the 103 or 97% are within 2 standard deviations. 101 of the 103 or 98% are within 3 standard deviations. This data seems to fit pretty well with the Empirical Rule.

2.99   **a.** A 52-centimeter head circumference will not occur often, but it will occur. The value 52 is 2 standard deviations below the mean ($z = \dfrac{\text{value - mean}}{\text{s.d.}} = \dfrac{52-56}{2} = -2$). This is at the lower end of the interval that describes about 95% of the values. Thus only about 2.5% of male head circumferences are smaller.
        a. A 62-centimeter head circumference will be rare. The value 62 is 3 standard deviations above the mean ($z = \dfrac{\text{value - mean}}{\text{s.d.}} = \dfrac{62-56}{2} = 3$). This is at the upper end of the interval that describes about 99.7% of the values. Thus only about 0.15% (about 3 in 2000 men) will have a larger circumference.

2.101   **a.** About 68% fall in the interval 540 ± 50, which is 490 to 590.
          About 95% fall in the interval $540 \pm (2)(50)$, which is 440 to 640.
          About 99.7% fall in the interval $540 \pm (3)(50)$, which is 390 to 690.

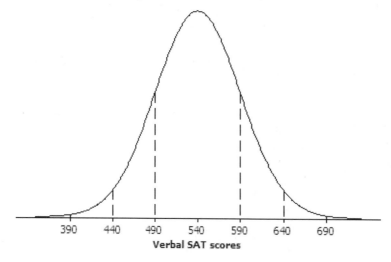

Figure for Exercise 2.101a

        b. $s^2 = 50^2 = 2500$ (Variance = squared standard deviation.)

2.103   **a.** 590, which is 1 standard deviation above the mean.
        b. 640, which is 2 standard deviations above the mean.

12

c. 490, which 1 standard deviation below the mean.

2.105 A categorical variable cannot have a bell-shaped distribution. A variable must be quantitative for it to be possible to have a distribution with any particular shape. For a categorical variable, the raw data are category labels without a meaningful numerical ordering.

2.107 **a.** Sleep, Dad's Height, Ideal Height for females, and Handspan for males are reasonably well described by the Empirical Rule, as is Handspan for males if the outlier is ignored. The variables TV, Exercise, and Alcohol are not well described by the Empirical Rule.
b. Variables well described by the Empirical Rule have roughly a bell shape. Variables not well described by the Empirical Rule have a skewed shape.

2.109 **a.** The shape is skewed, which makes it hard to judge outliers as the extreme points may just be part of the skewed pattern.
b. The Empirical Rule will not hold.
c. The interval is −3.863 to 12.08, and 90.42% of the data values are in this interval, decidedly more than the 68% that would be expected if the Empirical Rule applies.
d. The interval is −11.83 to 20.05. 97.01% of the data values are in this interval, a bit more than the 95% expected if the Empirical Rule applies. But, the interval contains negative values, which are impossible for this variable (number of drinks per week).

2.111 **a.** Outlier is at 100 hours. The remaining data is skewed so the Empirical Rule would not hold.
b. Outliers occur at 13 and 14 cm. The remaining data is roughly bell-shaped so the Empirical Rule may hold.
c. Outlier is at 55 inches. The remaining data is roughly bell-shaped so the Empirical Rule may hold.

2.113 **a.** The lower quartile Q1 = 0 hours. By definition, 25% of the values are at or below the lower quartile.
b. 0 to 2 hours (Minimum to Median).
c. 2 to 70 (Median to Maximum).
d. Yes, 70 hours would be marked as an outlier.
The boundary defining an outlier on the high side is $Q3 + 1.5 \times IQR = 5 + 1.5\,(5\text{-}0) = 12.5$ hours.
e. Range/6 = (70-0)/6 = 11.67.This is notably greater than the standard deviation. The outlier (70 hours) and skewness in the data cause Range/6 to differ from the standard deviation.
f. The mean is greater than the median. The outlier and skewness to the right causes this to occur.

2.115 **a.** Telephone exchange is a categorical variable.
b. Number of telephones is a quantitative variable.
c. Dollar amount of last month's phone bill is a quantitative variable
d. Long distance phone company used is a categorical variable.

2.117 **a.** Yes, a variable can be both explanatory and categorical. The phrase "explanatory variable" means that the variable might influence a response variable, and there is no restriction concerning whether the explanatory variable is categorical or quantitative (or ordinal). For an example, consider Example 2.2 in which type of nighttime lighting is a categorical explanatory variable.
b. No, a variable cannot be both continuous and ordinal. The term "ordinal variable" is used when the raw data are ordered categories while "continuous" means that all values in an interval are possible.
Supplemental note: Some might ask whether the term "ordinal" could apply to a continuous variable because the raw data can be used to order the sample observations. A restriction of ordinal numbers, however, is that all possible values can be counted. For a continuous variable, however, all possible values in an interval cannot be counted. There always are an infinite number of values between any two points in the interval so it's impossible to determine what "exact" value is second, third, and so on.
c. Yes, a variable can be both quantitative and a response variable. Generally, there is no restriction concerning whether a response variable is quantitative, categorical, or ordinal. For an example, consider Example 2.5 in which the quantitative response variable is right handspan (and the explanatory variable is sex).

d. Yes, a variable can be both bell-shaped and a response variable. An example is verbal SAT, which is designed to have bell-shaped distribution, and would be a response variable in a comparison of the verbal SAT scores of females and males.

2.119 **a.** The mean will be larger than the median. While most households may have between 0 and 4 or so children, there will be some households with large numbers of children, so the distribution will be skewed to the right.
b. The mean will be larger than the median. People like Bill Gates will create large outliers. And, generally income data tends to be skewed to the right because high incomes can become quite high but incomes can't be any lower than 0.
c. If all of the high school students are included, the mean will be higher than the median. This is because many high school students are too young to work or do not want to work, resulting in many students with $0 income earned in a job outside the home. There is even a chance the median could be 0!
d. The mean is 10.33 cents. Calculate this assuming there is one of each type of coin. The calculation is $(1+5+25)/3 = 31/3 = 10.33$. The exact number of each type of coin doesn't matter. As long as there are equal numbers of each type, the mean will be 10.33 cents. The median is the middle amount so it will be 5 cents. The mean is higher than the median because the monetary amounts are skewed to the right.

2.121 The interval is −6.1 to 22.7 hours. The interval includes negative values, which are impossible times. Thus, the interval based an assumption of a bell-shaped curve would not reflect reality.

2.123 **a.** The First Ladies may constitute a population rather than a sample. They lived in unique circumstances, so it is hard to view these women as a representative sample from any larger population. And, they can't be considered to be a sample from a larger population of First Ladies because future First Ladies will have different circumstances affecting life expectancy.
b. If the First Ladies are viewed as a population, the population standard deviation is $\sigma = 15.13$ years. In Excel, this can be found with the command "=STDEVP( )" and many calculators have a key for the population standard deviation. If the argument is made in part (a) that the First Ladies constitute a sample, the correct answer here is that the sample standard deviation is s 15.34 years.

2.125 **a.** Amount of beer consumed is the explanatory variable. Systolic blood pressure is the response variable.
b. Daily caloric intake of protein is the explanatory variable. Presence of colon cancer is the response variable.

2.127 **a.** If the two possible outliers are ignored, the data appear to be more or less bell-shaped so the Empirical Rule may hold.
b. The Empirical Rule implies that the range should span about 4 to 6 standard deviations. About 95% of the data will be within 2 standard deviations (plus or minus) of the mean and about 99.7% of a data set should be within 3 standard deviations (plus or minus) of the mean. Here, *range = maximum − minimum* = $23.25 - 12.5 = 10.75$ cm. This span is equal to $10.75/1.8 = 5.97$ standard deviations so it is consistent with the Empirical Rule.

2.129 **a.** The mean is 57.84 years and the sample standard deviation is $s = 6.997$ years.
Note: If this batch of data is viewed as a population rather than a sample, the standard deviation is 6.926 years. See the "technical note" in section 2.7 for an explanation of the difference between sample and population standard deviations.
b. *Range = maximum − minimum* = $(78 - 42) = 36$ years. This is a span of $42/6.997 = 5.14$ standard deviations, so the stated relationship between range and standard deviation does hold for the data.
c. For the youngest CEO, $z = \dfrac{\text{observed value} - \text{mean}}{\text{standard deviation}} = \dfrac{42 - 57.84}{6.997} = -2.264$.

For the oldest CEO, $z = \dfrac{74 - 51.47}{6.997} = 2.881$.

Remember that a *z*-score measures the number of standard deviations a value is from the mean. These values are about what you would expect because the Empirical Rule states that about 95% of the values fall within 2 standard deviations of the mean, and 99.7% of the values fall within 3 standard deviations of the mean. So, *z*-scores for the lowest and highest values are often somewhere between 2 and 3 in absolute magnitude.

2.131   What percentage of kindergarten children lives with their mother only? Their father only? One or both grandparents?

2.133   Is the average amount of coffee consumed per day the same for married people as it is for single people?

2.135   **a.** Low = 0, Q_1= 22.5, median = 55, Q_3 = 175, High = 450. To find these values, first write the data in order. The median is the average of the middle two values. The lower quartile is the median of the smallest 12 values and the upper quartile is the median of the larger 12 values.
b. 450 would be marked as an outlier. The boundary for upper outliers is 175 + 1.5 (175 − 22.5) = 403.75.

2.137   Low = 109, Q_1 =180.75, Median = 187.25, Q3 = 199, High = 214; 10; 109 is an outlier

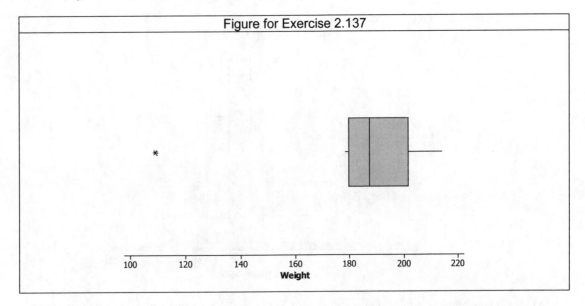

Figure for Exercise 2.137

2.139   Outliers affect the standard deviation. This happens because the calculation uses the deviation from the mean for every value. An outlier has a large deviation from the mean, so it inflates the standard deviation. Extreme values generally do not affect the quartiles, and consequently they generally don't affect the interquartile range. Remember that a quartile is determined by counting through the ordered data to a particular location, so the exact size of the largest or smallest observations doesn't matter.

2.141   **a.** The boxplot does not show the bimodal nature of the distribution. The figure for this exercise is on the next page

15

Figure for Exercise 2.141a

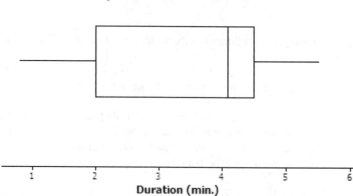

Duration (min.)

b. The distribution is bimodal.

Figure for Exercise 2.141b

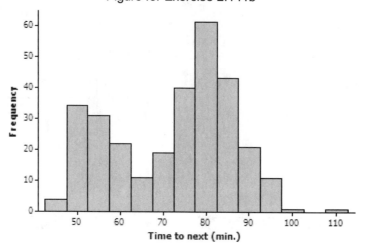

Time to next (min.)

2.143   **a.** Generally, the distribution is bell-shaped, although there are two outliers at 16 hours of sleep. Disregarding the outliers, the center of the distribution is somewhere near 7 hours.

Exercise for Exercise 2.143a

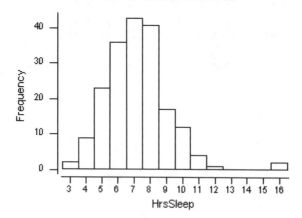

HrsSleep

16

b. The five-number summary is:

	Hours of sleep	
Median	7	
Quartiles	6	8
Extremes	3	16

c. *Range = maximum−minimum =* 16−3 = 13 hours. $IQR = Q_3 − Q_1 =$ 8−6 = 2 hours.

2.145   **a.** Of 1,902m respondents who answered this question, 1263/1902=66.4% said they favor capital punishment and 639/1902 = 33.60% said they oppose it. The following output was created using **Stat>Tables>Tally individual variables** in Minitab version 14.

Output for Exercise 2.145a		
cappun	Count	Percent
Favor	1263	66.40
Oppose	639	33.60
N=	1902	
*=	121	

b. The sample size is different for this part than for part (a) because nine people who answered the question about capital punishment did not give a political party preference. Useful conditional percentages are the percentages within the various political preferences. The counts and percentages determined using **Stat>Tables>Cross Tabulation** in Minitab, are:

	Favor	Oppose	All
Democrat	372 (53.9%)	318 (46.1%)	690
Independent	457 (67.3%)	222 (32.7%)	679
Other	22 (61.1%)	14 (38.9%)	36
Republican	406 (83.2%)	82 (16.8%)	488
All	1257 (66.4%)	636 (33.6%)	1893

c. The variables are related. The percentage in favor of capital punishment is noticeably less for Democrats than it is for Republicans. For Independents, the percentage in favor is between what it is for Democrats and Republicans.

2.147   **a.**

Figure for Exercise 2.147a

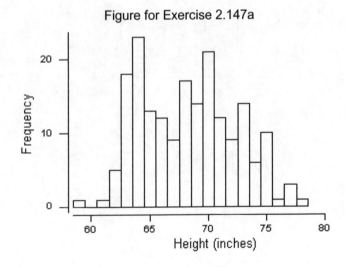

17

b. It's difficult to describe the shape of the histogram, but it's easy to say what it is not. It's not bell-shaped, and it's not skewed. Theoretically, it is bimodal, which means that there are two different peaks. In this case one peak occurs near the mean height of women and the other occurs near the mean height for men.

c.

Figure for Exercise 2.147c

Height (inches)

d. The histogram is more informative because it gives more detail about the pattern between 64 inches and 71 inches. We're able to see the two distinct peaks in the histogram.

18

CHAPTER 3
ODD-NUMBERED SOLUTIONS

3.1 **a.** Negative, because coordination will decrease when amount of alcohol consumed is increased.
b. No association would be expected between height and grade point average.
c. Negative, because heavier cars use more gas and can go fewer miles on a gallon of gas.

3.3 **a.** Negative association. As percent taking the test increases, average math SAT decreases.
b. More or less linear. There could be a slight curvature but a straight line appears to be a suitable description of the pattern.
c. Highest math Sat was around 600; about 5%, perhaps fewer, took the test in that state.
d. Lowest math SAT was around 475; about 60% took the test in that state.

3.5 **a.** Appropriate because both variables are quantitative.
b. Not appropriate because both variables are categorical.

3.7 **a.** The speed of the car is the explanatory variable and stopping distance is the response variable.
b. See the scatterplot below. There is a positive association that appears to be curvilinear.

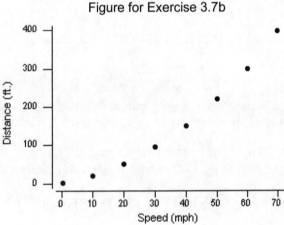

Figure for Exercise 3.7b

3.9 **a.**

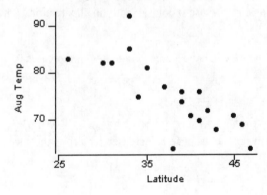

Figure for Exercise 3.9a

b. The pattern appears to be linear and the direction is negative.
c. Phoenix may be an outlier because its August temperature (92° F) is high for its latitude (33). San Francisco may be an outlier because its August temperature (64° F) is low for its latitude (38). Miami may

19

be an outlier because its August temperature (83) is low for its latitude (26). Both Miami and San Francisco may benefit from their proximity to water.

3.11   **a.** Asking price is the response variable and square footage is the explanatory variable.
     b. The scatter plot is shown below.

Figure for Exercise 3.11b

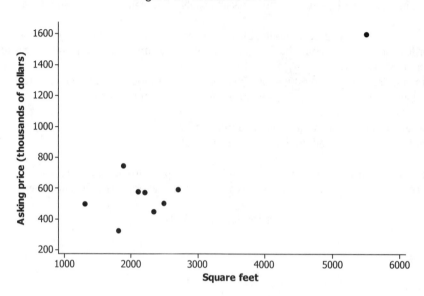

     c. "The individual in question belongs to a different group than the bulk of individuals measured." The other houses probably are in suburban neighborhoods, while the outlier is a beachfront mansion.

3.13   **a.** Average weight = –250 + 6(70) = –250 + 420 = 170 lbs.
     b. The slope is 6. On average weight increases 6 pounds for each 1-inch increase in height.

3.15   **a.** Average math SAT decreases 1.11 points per each 1-percent increase in the percentage of high school graduates taking the test.
     b. Average Math = 575 – 1.11(8) = 575 – 8.88 = 566.12.
     c. Residual = Actual y – Predicted $y = y - \hat{y} = 573 - 566.12 = 6.88$.

3.17   This is a deterministic relationship because it holds exactly in all situations. There is never variation from the given equation.

3.19   **a.** For each one-centimeter increase in handspan, average height increases 0.7 inches.
     b. $\hat{y} = 51.1 + 0.7(20) = 65.1\,\text{in.}$
     c. $66.5 - 65.1 = 1.4\,\text{in.}$

3.21   **a.** You would expect a success rate of 76.5 – 3.95(10) = 37%.
     b. The slope of –3.95 shows when the distance is increased by a foot, the success rate decreases by about 3.95%., on average.

3.23   **a.** $\hat{y} = 17.8 + 0.894(70) = 80.38$
     b. Residual = Actual y – Predicted y = 76 – 80.38 = –4.38

20

3.25 **a.** Here are the values of $\hat{y}$ and the values of $y - \hat{y}$ for Line 1:

x	1	2	3	4	
y	4	10	14	16	
$\hat{y}$	6	9	12	15	
$y - \hat{y}$	–2	1	2	1	

Therefore, for Line 1, $SSE = (4-6)^2 + (10-9)^2 + (14-12)^2 + (16-15)^2 = 10$.

Using similar arithmetic for Line 2, $SSE = (4-5)^2 + (10-9)^2 + (14-13)^2 + (16-17)^2 = 4$.

b. Line 2 is better because it has a smaller sum of squared errors than Line 1 does.

3.27 –1.7 and 2.5. They are not between –1 and +1.

3.29 The two variables are not linearly related. A scatter plot may show an association of some type, but it would not be a linear relationship.

3.31 **a.** As length increases, chest girth increases. The correlation value (r = 0.82) indicates a positive association.

b. The value 0.82 is indicates a strong association. The closer the correlation value is to 1 (in absolute value), the stronger the association.

c. The correlation would still be 0.82. The value of correlation is not changed if either or both variables are rescaled to different units.

3.33 The correlation would still be 0.95. A correlation value doesn't change when the measurement units of one or both of the variables are changed.

3.35 The correlation must be 1.0.

3.37 Graph 2 shows the strongest relationship while Graph 3 shows the weakest.

3.39 A correlation of $r = +0.12$ is low. In the sample, age and hours of watching television per day have a very weak, positive association.

3.41 **a.** Any of the three choices could be justified. Perhaps the correlation is negative because better students might watch less TV. Or perhaps it is positive because students with no social life tend to study more and watch more TV than students who are busy doing other things. In student surveys done by the authors of this text the correlation is close to 0, and a correlation close to 0 could also be justified by arguing that the two should have little to do with each other. (For instance, in the dataset **UCDavis1** on the companion website, the correlation between these two variables is −0.05.)

b. Positive because big cities will have many of each and small towns will have few of each.

c. Negative because strength decreases as age increases.

3.43 **a.** $r^2 = (0.4)^2 = .16$. This means that height explains 16% of the variation in weight.

b. The correlation would still be 0.40. Changing the units of measurement does not change the correlation.

3.45 $r^2 = (-0.36)^2 = .1296$ or about 13%. So, varying hours of study explains about 13% of the observed variation in hours of sleep.

3.47 $r^2 = \dfrac{SSTO - SSE}{SSTO} = \dfrac{800 - 200}{800} = 0.75$.

3.49 We know there is a positive association, but do not know the strength of the association.

3.51 It is not a reasonable assumption. In general it is not wise to extrapolate very far beyond the range of data available, with the assumption that a present trend will continue indefinitely.

3.53 Answers will vary, but here are two examples illustrating an outlier that deflates a correlation. In each case, the arrow points to the outlier.

Figure for Exercise 3.53

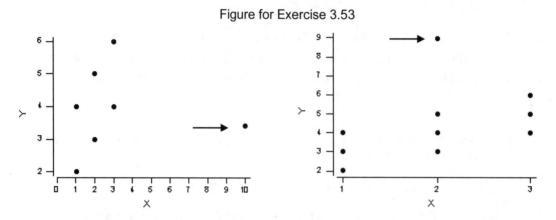

3.55 **a.** Estimated stopping distance when speed is 80 miles per hour is −44.2 + 5.7(80) = 411.8 ft. This is probably not an accurate estimate. A scatterplot shows that the relationship is likely not linear.
b. The scatter plot is shown in the answer for Exercise 3.7(b). From the plot, the relationship looks curvilinear. Extending the pattern of the scatterplot leads to an estimate of a little more than 500 feet.
c. Vehicle weight, road condition (wet or not), type of braking system, and driver pressure applied to the brakes are among the factors likely to affect braking distance in practice. The given data will not hold exactly in most situations.

3.57 The negative correlation might occur due to inappropriately combining groups (males and females). Perhaps females, who are generally shorter than males, did better on the memorization test. For the data in the following example sketch, the overall correlation is −0.685, but within each sex the correlation is close to 0.

Figure for Exercise 3.57

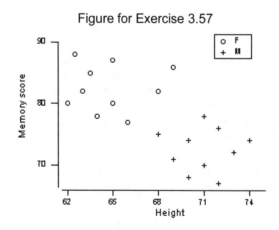

3.59 The answer will vary. Remember that an extrapolation is a prediction of a value well outside the range of observed data. An example is x = year and y = number of cell phones sold that year. The relationship based on data from the recent past will not hold into the distance future.

3.61 **a.** At 2 ft., the predicted success rate is 76.5 − 3.95(2) = 68.6%, well below the observed success rate of 93.3% (given in Part (b)). At 20 ft., the predicted success rate is 76.5−3.95(20) = −2.5%, much different than the observed rate of 15.8% (given in Part (b)) and also an impossible value because a success rate cannot be negative.

b. The equation predicts success rates that are much too low in both cases, indicating that the true relationship is probably not linear much beyond the range of 5 feet to 15 feet. This illustrates why it is not a good idea to use the regression relationship to predict *y* values beyond the range of the observed *x* values.

c. Based on the information provided, the relationship for the entire range from 2 ft. to 20 ft. is likely to be similar to this:

Figure for Exercise 3.61c

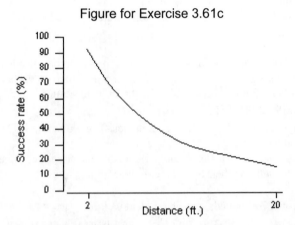

3.63   Sales of both will tend to be high in the colder months and low in the warmer months.

3.65   Sex of the respondents can explain the negative correlation. Females generally have more ear pierces and tend to be shorter than males. So, over the whole dataset, it will appear that as height increases the number of ear pierces decreases.
Note: With statistical software, it can be found that for males, the correlation between height and ear pierces is −0.022. For females, the correlation is −0.014.

3.67   **a.** One possibility is Reason #2 given in Section 3.5: "There may be causation, but confounding factors make this causation difficult to prove." In this case, it may be that people who smoke more may also drink more alcohol, exercise less, and have other behaviors that are likely to lower the age of death.
b. The most likely explanation is Reason #3 given in Section 3.5: "The observed association can be explained by how one or more other variables affect both the explanatory and response variables." In this case, both variables are related to the number of people at the ski resort on a given day.

3.69   The answers will vary. A randomized experiment gives the strongest evidence of a cause and effect relationship, so a good strategy for answering this question is to describe a randomized experiment. An example is to randomly assign volunteers to drink differing amounts of caffeine, and measure their heart rate 10 minutes later. If a positive correlation were found, it is likely that increased caffeine caused the increase in heart rate.

3.71   Answers will vary. As an example, yearly amount of travel on highways in a country and yearly sales of shoes in the country are likely to be increasing over time due to increasing population size.

3.73   No, remember that correlation does not prove causation. It's possible that the association may be due to others factors that are correlated with the dietary fat intake in countries and the development of breast cancer. Possible confounding variables include alcohol use, higher sugar content in the diet, lack of exercise and so on. Any variable that is systematically different in countries with high-fat and low-fat diets could be contributing to the observed relationship

3.75   Answer will vary, but an example is a plot that looks approximately like Figure 3.12 in the text.

3.77   Answer will vary, but the best line through the points will be flat. The plots in Figure 3.20 are examples.

3.79   Try putting a rectangular grid of points in the upper left corner of the plot. Put the outlier in the lower right corner.

3.81   **a.** Body temperature is the response variable and age is the explanatory variable.

b. $\hat{y} = 98.6 - 0.0138x = 98.6 - 0.0138(50) = 97.91$ degrees.

c. The predicted temperature for a person whose age = 50 was found in part (b) to be 97.91. Residual = actual – predicted = $97.6 - 97.91 = -0.31$. His body temperature is 0.31 degrees lower than would be predicted based on his age.

3.83   **a.** The intercept is19.42. It does not have a meaningful interpretation because height cannot be 0.

b. The slope is 0.658, indicating that if one woman's father was 1 inch taller than another woman's father, the one with the taller father would be predicted to be 0.658 inches taller.

c. $\hat{y} = 19.42 + 0.658x = 19.42 + 0.658(70) = 65.48$ inches

d. Residual = actual – predicted = $67 - 65.48 = 1.52$ inches. The woman is 1.52 inches taller than would be predicted based on her father's height.

e. No. The relationship is based on female students. Male students are taller in general, and a different relationship would hold for relating their heights with their fathers' heights.

3.85   **a.** For a woman who weighs 140 pounds, predicted ideal weight is $\hat{y} = 44 + 0.6(140) = 128$ lbs. For a man who weighs 140 pounds, predicted ideal weight is $\hat{y} = 53 + 0.7(140) = 151$ lbs. The predictions indicate that the average woman who weighs 140 lbs. would like to weigh 12 lbs. less, while the average man who weighs 140 lbs. would like to weigh 11 lbs. more.

b. No, the intercepts do not have logical physical interpretations in this example because people cannot weigh 0 pounds. Remember that an intercept gives the value of y when $x = 0$.

c. Yes, the slopes have a logical interpretation. For each sex, the slope gives the increase in reported ideal weight for each one pound increase in actual weight, which is 0.6 pound for women and 0.7 pound for men.

3.87   **a.** Average foot length increases 0.384 centimeters for each one-inch increase in height. This is the slope of the line.

b. The predicted difference in foot lengths is (10)(0.384)=3.84 cm.

c. The predicted foot length is $\hat{y} = 0.25 + 0.384(70) = 27.13$ cm., and the prediction error (residual) is $28.5 - 27.13 = 1.37$ cm.

3.89   **a.** The scatterplot is shown below.

Figure for Exercise 3.89

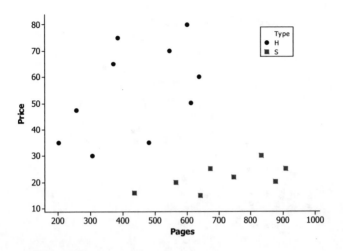

b. The correlation between price and pages is −0.309.
c. For hardcover books, the correlation is $r = 0.497$. For softcover books, the correlation is $r = 0.626$.
d. Inappropriately combining groups.

3.91   **a.** The regression equation is $\hat{y} = 44.4 - 0.0209 \times \text{year}$. For 2010, the estimated value is
$44.4 - 0.0209(2010) = 2.391$ persons per household.

b. The slope is −0.0209, which means that every year the average persons per household decreases by 0.0209, or about 0.2 persons per 10 year census cycle.

c. For 2200, the estimated value is $44.4 - 0.0209(2200) = -1.58$, which is an impossible value because 1 is the lowest possible number of persons per household (assuming a household is defined to be a place where somebody lives).

d. The Figure for Exercise 3.90 part (a) gives the pattern so far, and it looks like it may already be starting to flatten out in the year 2000. In the future, the pattern will flatten out and perhaps even increase.

3.93   Answers will vary, but one example is shoe size and head circumference. Neither causes the other to be larger, but they are both related to the overall size of a person

3.95   **a.** Determine the nature of the relationship.
b. Predict college grade point average in the future, based on SAT score.
c. Predict future height at age 21, based on height at age 4.
d. Determine the relationship

3.97   **a.** There is a positive linear association between the two variables. There is a moderately strong association. There are no notable outliers. Figure for this exercise is on the next page.

25

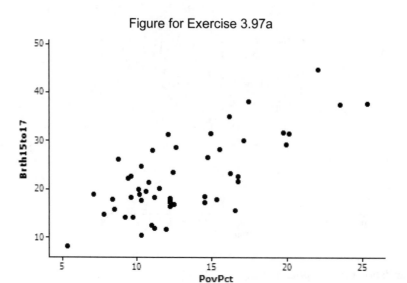

Figure for Exercise 3.97a

b. $\hat{y} = 4.267 + 1.373x$, where y = birth rate for females 15 to 17 year old and x = poverty rate (as a percent).

c. The slope is 1.373. On average, the birth rate (per 1000 persons) for females 15 to 17 years old increases by 1.373 for each one-percent increase in the poverty rate.

d. Predicted birth rate = $\hat{y} = 4.267 + 1.373(15) = 24.86$.

3.99 **a.** There is a moderately strong positive association between the variables. There are no data points that have a combination of values that are not consistent with the overall pattern. Some may note the gap between the bulk of the data and the individual with the highest cholesterol values.

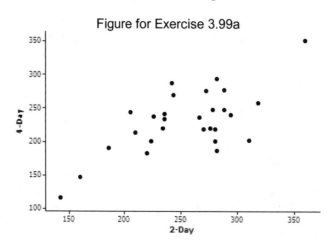

Figure for Exercise 3.99a

b. $\hat{y} = 62.37 + 0.6627x$, where y= 4-day measurement and x = 2-day measurement.

c. Slope = 0.6627. Average 4-day measurement increases 0.6627 per each 1-unit increase in the 2-day measurement.

d. With 2-day = 200, predicted 4-day = $62.37 + 0.6627(200) = 194.90$ With 2-day = 250, predicted 4-day $= 62.37 + 0.6627(200) = 228.04$. With 2-day = 300, predicted 4-day $= 62.37 + 0.6627(200) = 261.17$.

e. The difference between the 2-day and the 4-day measurements becomes greater as the 2-day measurement increases.

3.101 **a.** There is a strong positive association with a linear pattern.

26

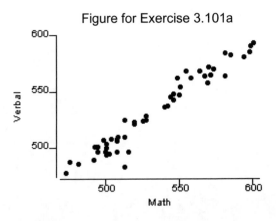

Figure for Exercise 3.101a

b. Hawaii may be an outlier because its average verbal SAT (483) is low relative to its math average (513). (Additional suggestion: calculate the difference between the two scores for all states and then draw either a dotplot or boxplot of those differences. Hawaii will be an obvious outlier.)

3.103   **a.** There is a positive association with a linear pattern. The difference increases, on average, as actual weight increases. Note that a positive difference occurs when actual weight is more than ideal weight.

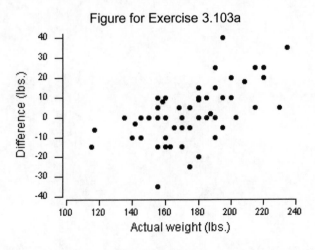

Figure for Exercise 3.103a

        b. The regression equation is $diff = -52.5 + 0.312 \, Actual$. For a 150-pound man, the predicted difference is $-52.5 + 0.312(150) = -5.7$ lbs., which means that actual weight is 5.7 pounds less than ideal weight. On average, 150-pound men want to weight more than they do.

        c. The predicted difference is $-52.5 + 0.312(200) = 9.9$ lbs., which means that actual weight is 9.9 pounds greater than ideal weight. On average, 200-pound men want to weigh less than they do.

        d. $r^2 = .353$, or 35.3%.

3.105   **a.** The regression equation is $height = 24.7 + 0.6 \, midparent$.

        b. $\hat{y} = 24.7 + 0.6(68) = 65.5$ in.

        c. The mid-parent height is (70+62)/2 = 66 in., so $\hat{y} = 24.7 + 0.6(66) = 64.3$ in.

        d. A scatterplot and a correlation (or r^2).

CHAPTER 4
ODD-NUMBERED SOLUTIONS

4.1　**a.** $(1357/2057) \times 100\% = 64.3\%$. This is a row percentage. It is calculated as the percentage of the total number in a row of the table.

b. $(83/548) \times 100\% = 15.1\%$. This is a column percentage. It is calculated as the percentage of the total number in a column of the table.

c. $(285/2858) \times 100\% = 10.0\%$

d.

	Sunscreen Use		
Grade	*Never or rarely*	*Sometimes*	*Always or Most Times*
As and Bs	64.3%	21.9%	13.9%
Cs	81.4%	11.9%	6.7%
Ds and Fs	82.5%	14.6%	2.9%

e. There looks to be a relationship. The percentage that never or rarely wears sunscreen is greater for students who typically get Cs or Ds and Fs than it is for students who typically get As and Bs. Correspondingly, the percentage that wears sunscreen always or most times is higher for students who get As and Bs.

4.3　**a.** $(170/270) \times 100\% = 63.0\%$. This means that of the 270 freshman who auditioned, 63.0% were females.

b. $(170/300) \times 100\% = 56.7\%$. This means that of the 300 females who auditioned, 56.7% were freshman.

c. Freshman had 63.0%; Sophomores had 50.0%; Juniors had 75.0%; Seniors had 40.0%. Thus, juniors had the highest percentage of female applicants.

d. Of females, $(50/300) \times 100\% = 16.7\%$ were sophomores; Of males, $(50/200) \times 100\% = 25.0\%$ were sophomores. Males had a higher percentage of sophomore applicants.

4.5　**a.** Yes, both variables are categorical.

b. No, both variables are quantitative.

4.7　No, it is not sufficient. Totals are given for categories of each variable but counts for combinations are not provided. We do not know how many of the men were Democrats and how many were Republicans nor do we know how many women were Democrats and how many were Republicans

4.9　**a.**

	Prefer take-home	*Prefer in-class*	*Total*
A on midterm	10	15	25
Not A on midterm	30	20	50
Total	40	35	75

b. The response variable is preference for type of final exam. The explanatory variable is grade on the midterm.

c. Among students who got an A on the midterm, $(10/25) \times 100\% = 40\%$ prefer a take-home final exam (and 60% prefer an in-class exam).

Among students who did not get an A on the midterm, $(30/50) \times 100\% = 60\%$ prefer a take-home final exam (and 40% prefer an in-class exam)

There is relationship between the two variables. Students who did not get an A on the midterm are more likely to prefer a take-home final exam than are the students who got an A on the final.

4.11　**a.** Percentage ever bullied $= (42/92) \times 100\% = 45.65\%$ among short students

b. Percentage ever bullied = $(30/117) \times 100\% = 25.64\%$ among students not short

c. Yes, there is a relationship. Short students are more likely to have been bullied than students who are not short.

4.13   There are only minor differences between the percentage distributions given for men and women. Thus it does not seem that gender and reason for taking care of your body are related variables.

4.15   **a.** 1.0. This occurs when the risk is the same in each category.
   b. 1
   c. 0%

4.17   **a.** Drug 1: 10/100 = .10 (or 10%); drug 2: 5/100 = .05 (or 5%)

   b. Relative risk =Risk in drug group/Risk in placebo group = $\dfrac{10/100}{5/100} = 2.$

   c. Percent increase = $(2 - 1) \times 100\% = 100\%$, computed as (Relative risk $-1)\times100\%$. Relative risk is in solution for part (b). Could also be computed as $\dfrac{.10-.05}{.05} \times 100\%$.

   d. Odds ratio = $\dfrac{10/90}{5/95} = 2.11.$

4.19 $5 \times 1\% = 5$ (or .05 when expressed as a proportion).

4.21   **a.** $(1.53 - 1) \times 100\% = 53\%$
   b. 2.40. To find this, start with the equation *Increase in risk* = (*Relative risk* $- 1) \times 100\%$, substitute 140% for *Increase in risk*, and solve for *Relative risk*.

4.23   No, we would need to know the baseline risks for males and females.

4.25   **a.** Relative risk $=\dfrac{\text{Risk for "most anger" group}}{\text{Risk for "no anger" group}}=\dfrac{59/559}{8/199}=\dfrac{.106}{.04}=2.65$

   Equivalently, the percentages determined for Exercise 6.9 parts (b) and (c) can be used. If so, the calculation is 10.6% / 4% = 2.65.

   b. Percent increase in risk $=\dfrac{\text{Difference in risks}}{\text{Baseline risk}} \times 100\% = \dfrac{10.6 - 4}{4} \times 100\% = 165\%$.

   Equivalently, percent increase = (Relative risk$-1)\times100\%$ = (2.65$-1)\times100\%$ = 165%.

   c. $\dfrac{59/500}{8/191} = 2.82.$

4.27   **a.** For age 18 to 29, the percentage is $(212/1525) \times 100\% = 13.9\%$ who report having seen a ghost;
      For age 30 or over, the percentage is $(465/4377) \times 100\% = 10.6\%$ who report having seen a ghost.

   b. Relative risk $=\dfrac{\text{Risk for age 18 to 29}}{\text{Risk for age 30 or more}}=\dfrac{.139}{.106}=1.31$.

   Adults under 30 years old are 1.31 times as likely to report having seen a ghost as adults 30 years old or older.

   c. Increased risk = (Relative risk$-1)\times100\%$ = (1.31$-1)\times100\%$ = 31%

   Equivalently, $\dfrac{(\text{Risk for age 18 to 29}) - (\text{Risk for age over 30})}{\text{Risk for age over 30}} \times 100\% = \dfrac{.139-.106}{.106} \times 100\% = 31\%$

   Adults under 30 years old are 31% more likely to report seeing a ghost than adults over 30 years old.
   d. The odds of seeing a ghost are 465 to 3912, or about 1 to 8.4 (divide both counts by 465 to get the second set of odds given). In the older group 465 reported having seen a ghost and 3,912 said they had not. The odds compare these two values.

4.29 There may have been more homes in your town in 2005 than in 1990. The rate of burglaries per number of homes might the same or lower than in 1990.

4.31 The principal question to ask is, "What percentage usually fails?" In other words, what is the actual risk of failing? If it is small (say, 1%), then a doubled risk is not serious.

4.33 Probably not, although it's possible. For Simpson's Paradox to hold the relationship between blood pressure and religious activity would need to be reversed when separate categories of a third variable are considered.

4.35 Some possible "third variables" that could explain the difference are occupation, overall amount of long-distance driving, and amount of nighttime driving. Men may be more likely to have jobs that require long-distance driving. In general, men probably do more long-distance driving (leading to drowsiness at the wheel), and men probably do more nighttime driving.

4.37 **a.** The combined table is

	Admit	Deny	Total	Percentage Admitted
Men	450	550	1,000	450/1,000 = 45%
Women	175	325	500	175/500 = 35%
Total	625	875	1,500	

Of the men applicants, $(450/1000) \times 100\% = 45\%$ were admitted.
Of the women applicants, $(175/500) \times 100\% = 35\%$ were admitted.
Overall, men were more successful at gaining admission.
b. Program A admission rates: men, percentage is $(400/650) \times 100\% = 61.5\%$; women, percentage is $(50/75) \times 100\% = 67\%$. Program B admission rates: men, percentage is $(50/350) \times 100\% = 14.3\%$; women, percentage is $(125/425) \times 100\% = 29.4\%$.
In each program a higher percentage of women were admitted!
c. Simpson's Paradox occurs when combining groups reverses the direction of the relationship from what it was when the groups were separate, and this occurs in this situation. In both programs, the percentage of women applicants admitted was higher than the percentage of men applicants admitted. But, in the overall combined data, the percentage of women applicants admitted was lower than the percentage of men applicants admitted. Notice that the majority of men apply to Program A, which has a higher acceptance rate than program B. The overwhelming majority of the women apply to Program B which is tougher to get into than Program A, and this lowers the overall acceptance rate for women.

4.39 **a.** The response variable is outcome of treatment (successful or not).
b. In general, it seems likely that there will be a better success rate for less severely depressed patients than for more severely depressed patients. So, treatment B may have had a better success rate than expected because the two doctors assigned it to less severely depressed patients. The doctors probably thought they were doing a service to the more depressed patients by assigning them to the treatment that was originally thought to work better, but doing so probably produced misleading results.
c. Separate the patients into severity of depression groups. Then, compare the treatments within each separate group.

4.41 **a.**

Gender	Had Injury	No Injury	Total
Girls	74	925	999
Boys	153	1514	1667
Total	227	2439	2666

b.

Gender	Had Injury	No Injury	Total
Girls	$\dfrac{999 \times 227}{2666} = 85.1$	$\dfrac{999 \times 2439}{2666} = 913.9$	999
Boys	$\dfrac{1667 \times 227}{2666} = 141.9$	$\dfrac{1667 \times 2439}{2666} = 1525.1$	1667
Total	227	2439	2666

c. Chi-square = $\dfrac{(74-85.1)^2}{85.1} + \dfrac{(925-913.9)^2}{913.9} + \dfrac{(153-141.9)^2}{141.9} + \dfrac{(1514-1525.1)^2}{1525.1}$

$= 1.44 + 0.13 + 0.86 + 0.08 = 2.51$

4.43 **a.** Null is H_0: There is no relationship between gender and opinion on the death penalty.
Alternative is H_a: There is a relationship between gender and opinion on the death penalty.
b. Because the p-value is small (less than .0), we can conclude that in the larger population there is a relationship between gender and opinion on the death penalty.
c. For males, the percentage opposed = 28.7%, calculated as $(254/885)\times100\%$.   For females, the percentage opposed = 37.9%, calculated as $(385/1017)\times100\%$. The large difference in sample percentages is evidence that opposition to the death penalty is related to gender.

4.45 Null: No relationship between opinion about banning texting while driving and whether or not individual has texted while driving.
Alternative: There is a relationship between opinion about banning texting while driving and whether or not individual has texted while driving.

4.47 **a.** Reject the null hypothesis because p-value < .05.
b. Cannot reject the null hypothesis because p-value >0.05.
c. Reject the null hypothesis because p-value < .05.
d. Cannot reject the null hypothesis because chi-square < 3.84 (so p-value > .05).

4.49 No. For instance, with a very large sample size we may be able to declare statistical significance even for very weak relationships (small differences). The magnitude of the difference may not be important in a practical sense.

4.51 **a.** Yes. By the luck of random sampling, the observed sample might show a relationship when none actually exists in the larger population. We will learn in Chapters 12 and 13 that this type of result is called a Type 1 error.
b. Yes. By the luck of random sampling, the observed sample might not show a relationship when one actually exists in the larger population. We will learn in Chapters 12 and 13 that this type of result is called a Type 2 error, and the risk of such an error is influenced by the sample size.
c. No. Presuming that calculations for a chi-square test are done correctly, the analysis of an observed sample in which there's no relationship will give the conclusion that it is not a statistically significant relationship.

4.53 $\dfrac{(\text{Obs - Exp})^2}{\text{Exp}} = \dfrac{(212-174.93)^2}{174.93}$

4.55 **a.** Null hypothesis: There is no relationship between gender and opinion about capital punishment.

32

Alternative hypothesis: There is a relationship between gender and opinion about capital punishment.
b. The relationship is not statistically significant because p-value $=.19$ is greater than .05. The relationship in the sample did not provide enough evidence to conclude that there is a relationship in the population. Note that this is not the same as concluding that there is *no* relationship in the population. It could be that with a larger sample size, a significant relationship would be established.
c. Chi-square statistic $= 1.714$.
The contingency table of *observed counts* is:

	Favor	Oppose	Total
Men	38	12	50
Women	32	18	50
Total	70	30	100

The table of expected counts is:

	Favor	Oppose	Total
Men	$\dfrac{50 \times 70}{100} = 35$	$50 - 35 = 15$	50
Women	$70 - 35 = 35$	$50 - 35 = 15$	50
Total	70	30	100

The chi-squared statistic $=$

$$\frac{(38-35)^2}{35} + \frac{(12-15)^2}{15} + \frac{(32-35)^2}{35} + \frac{(18-15)^2}{15} = .257 + .600 + .257 + .600 = 1.714$$

Notice that for a two-way table, the formula $\dfrac{\text{Row total} \times \text{Column total}}{\text{Total n}}$ only has to be used to determine the expected count for one cell. Other expected counts can be determined utilizing the fact that the row and column totals for the expected counts are the same as they are for the observed counts.
c. The chance is .19 that we would get a chi-square value as large as (or larger than) 1.714 if there really is no relationship between the variables in the population.
d. The probability is .19 that the chi-square statistic would be as large as it is, or more so, if there is really no relationship in the population. Because the p-value is larger than .05 we cannot reject the null hypothesis of no relationship.

4.57   A small sample size provides limits our ability to make conclusions about a population. Failure to declare statistical significance is not absolute proof that there is no relationship in the population. A specific difference between observed conditional percentages can be statistically significant for a large sample size, but might not be statistically significant for a small sample size.

4.59   **a.** Null hypothesis: There is not a relationship between gender and opinion about legalization of marijuana. Alternative hypothesis: There is a relationship between gender and opinion about legalization of marijuana.
b. Because the p-value is less than .05, we are able to conclude that gender and opinion about legalization of marijuana are related variables.
c. For males, the percentage opposed $= 56.7\%$, calculated as $(315/556) \times 100\%$. For females, the percentage opposed $= 63.1\%$, calculated as $(436/691) \times 100\%$. The difference in sample percentages is evidence that opinion about legalization of marijuana may be related to gender.
d. The expected counts are:

	Legalize	Don't Legalize	Total
Male	$\dfrac{556 \times 496}{1247} = 221.15$	$556 - 221.15 = 334.85$	556
Female	$496 - 221.15 = 274.85$	$691 - 274.85 = 416.15$	691
Total	496	751	1247

$$\text{Chi-square} = \frac{(153-135.56)^2}{135.56} + \frac{(224-241.44)^2}{241.44} + \frac{(153-170.44)^2}{170.44} + \frac{(321-303.56)^2}{303.56} = 6.29.$$

4.61 **a.** Yes, there appears to be a relationship. As the number of ear pierces increases, the percentage with a tattoo also increases.

Pierces	% with Tattoo
2 or less	7.4% (40/538)
3 or 4	13.4% (58/432)
5 or 6	27.6% (77/279)
7 or more	42.1% (53/126)

b. First find that the total number for each column (no tattoo = 1147, tattoo = 228). Then, divide the counts by the appropriate column total to determine the column percentages. The graph shows that women who have a tattoo are more likely to have a large number of ear pierces. The percentages in the "5 or 6" and "7 or more" ear pierces categories are greater for the women who have a tattoo than they are for women who don't have a tattoo.

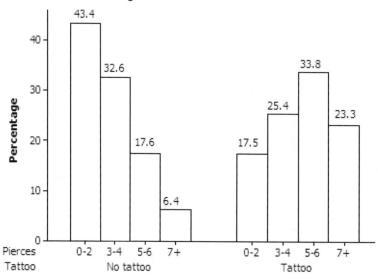

Figure for Exercise 4.61b

c. Among the 228 women with a tattoo, the number with 5 or more ear pierces is 77+53 = 130. The percentage is (120/228) × 100% = 57%.
Among the 1147 women who do not have a tattoo, the number with 5 or more ear pierces is 202+73 = 275
The percentage is (275/1147) × 100% = 24%.
d. Percentage with a tattoo = (228/1375) × 100% = 16.58%
e. Percentage with two or fewer pierces and no tattoo = (498/1375) × 100% = 36.22%.

4.63 **a.**

	Lung cancer	*Control*	*Total*
Owned bird	98	101	199
Never owned bird	141	328	469
Total	239	429	668

b. Bird owners: 98/199 = .4925, or 49.25% have cancer.
  Never owned bird: 141/469 =.301, or 30.1% have cancer.

34

c. No, these risks cannot be used as baseline risks. This was a case control study in which the researchers purposely sampled more lung cancer patients than would naturally occur in a group of 668 people. The sample percentages falling into the cancer category are much higher than you would see in a random sample.

d. Relative risk $= \dfrac{\text{Risk for bird owners}}{\text{Risk for those who never owned bird}} = \dfrac{.4925}{.301} = 1.64$.

e. You would want to know "baseline" risk of lung cancer for people like yourself (similar age, smoking habits, etc.). That risk is probably small.

f. No, a causal connection cannot be established. This is an observational study so there may be confounding factors that explain the results. The "case control" sampling took age and sex into account, but did not take other variables, such as smoking habits, into account.

g. Compare the smoking habits of the bird owners and non-owners to see whether or not the bird owners generally smoke more.

4.65   Chi-square = 4.817 and *p*-value = .028. The relationship is statistically significant because the *p*-value is less than 0.05. Minitab output is shown below (expected counts are shown below actual counts).

```
                    Minitab output for Exercise 4.65

                    First      Two +      All

    Smoker           29         71         100
                    38.74      61.26      100.00

    Nonsmoke        198        288         486
                   188.26     297.74      486.00

    All             227        359         586
                   227.00     359.00      586.00

    Chi-Square = 4.817, DF = 1, P-Value = 0.028

       Cell Contents --
                    Count
                    Exp Freq
```

The "by hand" calculation is as follows:
Compute the expected counts:

	First cycle	*2 or more cycles*	*Total*
Smoker	$\dfrac{100 \times 227}{586} = 38.74$	100–38.74 = 61.26	100
Nonsmoker	227-38.74 = 188.26	486–188.26 = 297.74	486
Total	227	359	586

Notice that for a two-way table, the formula $\dfrac{\text{Row total} \times \text{Column total}}{\text{Total n}}$ only has to be used to determine the expected count for one cell. Other expected counts can be determined utilizing the fact that the row and column totals for the expected counts are the same as they are for the observed counts.

Compute the chi-squared statistic:

$$= \frac{(29-38.74)^2}{38.74} + \frac{(71-61.26)^2}{61.26} + \frac{(198-188.26)^2}{188.26} + \frac{(288-297.74)^2}{297.74}$$
$$= 2.45 + 1.55 + 0.50 + 0.32 = 4.82$$

35

To determine the p-value: In Excel, use CHIDIST(4.82,1), to find p-value = .028. Also, we know the p-value is less than .05 because the value of the chi-square statistic is grater than 3.84.

4.67 **a.** $\dfrac{\text{Risk for hormone therapy group}}{\text{Risk for placebo group}} = \dfrac{166/8506}{124/8102} = \dfrac{.0195}{.0153} = 1.27$.

b. $(124/8102) = 0.0153$, or 1.53%.

c. The baseline risk (in conjunction with the relative risk) gives information about the actual risks involved. From the relative risk we would know only that the risk is 1.28 times greater for the hormone therapy group, but would not know the actual risk.

4.69 Because the p-value is .0005 (less than .05) we can say that there is a statistically significant relationship

4.71 Using the approximate odds given in Exercise 6.57, the odds ratio is $24/8.5 = 2.82$.

The precise odds ratio is found as Odds ratio $= \dfrac{\text{odds for "no anger"}}{\text{odds for "most anger"}} = \dfrac{191/8}{500/597} = \dfrac{23.875}{8.475} = 2.817$.

The odds of remaining free of heart disease versus getting heart disease for men with no anger are about 2.8 times the odds of those events for men with the most anger.

4.73 **a.** For men, the odds of being admitted to Program A are 400 to 250, or 1.6 to 1. (To get 1.6 to 1, divide both original counts by 250.) For women applying to Program A, the odds are 50 to 25, or 2 to 1.
The odds ratio for women compared to men is 2/1.6 or 1.25. The odds of being admitted for women are about 1.25 times what they are for men, so women have better odds.
b. For men, the odds of being admitted for the combined programs are 450 to 550, or about 0.8 to 1.
For women, the odds are 175 to 325, or about 0.54 to 1.
The odds ratio (women to men) is therefore about .54/.8 = .675. The odds of being admitted versus being denied admittance for women are about 2/3 of what they are for men, so women have worse odds of being admitted overall.
c. The last sentence of the part (a) solution makes it seem that the university favors women, while the last sentence of the part (b) solution makes it seem that the university favors males.

4.75 **a.**

	About right	*Overweight*	*Underweight*	*Total*
Female	109	32	7	148
Male	56	15	13	84

b.

	About right	*Overweight*	*Underweight*	*Total*
Female	73.65%	21.62%	4.73%	100%
Male	66.67%	17.86%	15.48%	100%

Of the men, 15.48% think they are underweight.
Of the women, only 4.73% think they are underweight.
c. The relationship is statistically significant. The p-value is .019 (chi-square statistic = 7.921). Minitab output is as follows:

36

```
                        Minitab output for Exercise 4.75c

                    AboutRt    OverWt    UnderWt     All

        Female        109        32         7        148
                    105.26     29.98     12.76     148.00

        Male           56        15        13         84
                     59.74     17.02      7.24      84.00

        All           165        47        20        232
                    165.00     47.00     20.00     232.00

        Chi-Square = 7.921, DF = 2, P-Value = 0.019

     Cell Contents --
            Count
            Exp Freq
```

d. The pattern is about the same at the two schools, although the percentage of men saying they are overweight is lower at Penn State than it is at UC Davis.

4.77   **a.** Null hypothesis: There is not a relationship between gun ownership (or not) and opinion about required police permits for guns.
Alternative hypothesis: There is a relationship between gun ownership (or not) and opinion about required police permits for guns.

b. The two-way table of counts, along with the percentage favoring and opposing permits for each gun ownership group, is:

	Favor permits	*Oppose permits*	*Total*
No gun in home	741 (86.36%)	117 (13.64%)	858
Yes, have gun in home	300 (66.65%)	157 (33.35%)	457
Total	1041	274	1315

In the whole sample, percentage owning a gun is $(457/1315) \times 100\% = 34.75\%$.

c. Gun in home: $(300/457) \times 100\% = 66.65\%$ favor permits
   No gun in home: $(741/858) \times 100\% = 86.36\%$ favor permits.
The difference in percentage favoring permits is $86.36 - 66.65 = 19.71\%$. This is evidence of a possible relationship between gun ownership and opinion about permits.

d. p-value ≈ 0. Yes, the relationship is statistically significant. Minitab output is as follows:

```
                    Minitab output for Exercise 4.77d
     Rows: owngun    Columns: gunlaw

              Favor   Oppose    All

     No         741      117    858
     Yes        300      157    457
     All       1041      274   1315

     Cell Contents:       Count

     Pearson Chi-Square = 77.594, DF = 1, P-Value = 0.000
```

e. An observed relationship is statistically significant if it is unlikely that a relationship as strong, or stronger, would be observed in a sample if there were no relationship in the population. In the context of this exercise, it is nearly impossible (p-value ≈ 0) that the difference in the sample percentages calculated in part (b) would be so large if there were no difference in opinion between those who have a gun and do not have a gun in the population.

4.79   **a.**

| Seat | Importance of Religion | | | |
	Very Important	Fairly Important	Not Important	Total
Front	45	63	43	151
Middle	87	199	118	404
Back	16	57	61	134

b.

| Seat | Importance of Religion | | | |
	Very Important	Fairly Important	Not Important	Total
Front	29.80%	41.72%	28.48%	100%
Middle	21.53%	49.26%	29.21%	100%
Back	11.94%	42.54%	45.52%	100%

c. As we move from the front to the back of the room students become less likely to say that religion is very important in their lives and more likely to say that religion is not important in their lives.

CHAPTER 5
ODD-NUMBERED SOLUTIONS

5.1 When the data can be considered to be representative of a much larger group with regard to the question(s) of interest.

5.3 **a.** Yes. Women in the psychology class probably will have heights similar to the heights of other women at the college.
b. No. Parents of children in daycare are likely to have different opinions than other adults in the state. They are likely to be more supportive of funding for daycare.

5.5 **a.** Population of interest = all registered voters in the community.
b. Sample = the 400 randomly selected individuals.

5.7 Selection bias occurs if the method for selecting the participants produces a sample that does not represent the population of interest.

5.9 This answer will differ for each student. There are many situations in which the available resources would not permit a census, and also there are many situations in which the process of taking the necessary measurements damages the product so a census would not be done. For instance, a manufacturer of potato chips would not test the quality of its product by tasting the contents of every bag it produced. They would select a sample to test instead.

5.11 A sample survey measures a subgroup of a population in order to learn something about that larger population, whereas a census measures everyone in the population.

5.13 No, this would not be a simple random sample because not all combinations of four songs are equally likely. Any set of four songs in which two or more are on the same CD would not have a chance of being the sample.

5.15 **a.** Response bias.
b. Selection bias.
c. Nonparticipation bias.

5.17 **a.** Selection bias would be introduced because the sample would not be representative of the population of interest. The sample would be representative of all registered automobile owners, not all homeowners.
b. Nonparticipation bias (nonresponse bias). People who feel strongly about the mayor's performance are more likely to respond.

5.19 **a.** No, the "Fundamental Rule for Using Data for Inference" does not hold. In professional basketball, women get paid much less than men. This is not representative of all careers is which men and women have equivalent jobs.
b. Yes, the "Fundamental Rule for Using Data for Inference" holds. The pulse rates of students in a statistics class at a large university are probably representative of the pulse rates of all college-age people.

5.21 Answers will vary. One type of example is a survey mailed to a random sample of a population, but for which only those with a strong opinion about the issue are likely to respond. For instance, a survey about the possibility of building a new sports stadium in a city is more likely to get responses from people who care about sports.

5.23 $\dfrac{1}{\sqrt{n}} = \dfrac{1}{\sqrt{5000}} = .014$. As a percentage, $.014 \times 100\% = 1.4\%$.

5.25 **a.** $\dfrac{1}{\sqrt{n}} = \dfrac{1}{\sqrt{90}} = .105$. As a percentage, $.105 \times 100\% = 10.5\%$.

b. $.800 \pm .105$, which is .695 to .905.

5.27 $\dfrac{1}{\sqrt{n}} = \dfrac{1}{\sqrt{2000}} = .022$ or 2.2%.

5.29 **a.** $\dfrac{1}{\sqrt{n}} = \dfrac{1}{\sqrt{1006}} = .032$. As a percentage, $.032 \times 100\% = 3.2\%$

b. Population proportion: $.40 \pm .032$ or .368 to .432. Percentage: $40\% \pm 3.2\%$ or 36.8% to 43.2%.

c. We have 95% confidence that the interval includes the percentage of all adult Americans thinking Internet shopping poses more of a threat. More specifically, for 95% of all intervals found in this way, the interval will cover the true population proportion.

d. Yes. The 95% confidence interval is 36.8% to 43.2%, an interval of values all less than 50%. It's reasonable to conclude that fewer than half of all adult Americans think Internet shopping poses more of a threat than buying by mail order or in a store.

5.31 **a.** About 95%, or 95 of the confidence intervals are likely to cover the true percent. Note that this does not mean that exactly 95 of them will cover the true percent. The fact that each poll has a 95% chance of covering the truth does not mean that exactly 95 of them will.

b. No, there is no way to know which of the polls cover the true percent.

5.33 The margin of error is .03 or 3%. The report says so with the statement "one can say with 95 percentage confidence that the maximum error attributable to sampling and other random effects is plus or minus 3 percentage points." Also, the margin of error can be computed from the given sample size

as $\dfrac{1}{\sqrt{n}} = \dfrac{1}{\sqrt{1021}} = .031$, or about 3%.

5.35 **a.** The sample proportion is .49 and the margin of error, expressed as a proportion, is .03. The interval, calculated as *sample proportion ± margin of error,* is $.49 \pm .03$, which is .46 to .52.

b. The sample proportion is .47 and the margin of error, expressed as a proportion, is .03. The interval, calculated as *sample proportion ± margin of error,* is $.47 \pm .03 = .44$ to .50.

c. No, we cannot conclude that a clear majority of the population has either opinion. Note that the intervals computed in the previous two parts both contain .50, so it's possible that about 50% of the population has each opinion.

5.37 **a.** $\dfrac{1}{\sqrt{n}} = \dfrac{1}{\sqrt{400}} = 0.05$ so the margin of error would be about 5%.

b. Table 5.1 shows that $n = 2500$ provides a margin of error of about 2%. To check this, calculate $\dfrac{1}{\sqrt{2500}}$. Or, solve for n in the formula $\dfrac{1}{\sqrt{n}} = .02$ to find $n = 2500$.

5.39 The margin of error will be smallest for sample B. As the sample size increases, the margin of error decreases. The population size does not affect margin of error (as long as the sample is a small part of the population).

5.41 At least $n = 400$. The answer can be found in Table 5.1 on page 153. Or, the answer can be calculated by solving for n in the equation $\dfrac{1}{\sqrt{n}} = .05$.

40

5.43   **a.** A probability sampling plan is any sampling method for which it is possible to specify the chance that any particular individual in the population will be selected for the sample.
b. A simple random sample is the result of a plan in which every conceivable group of units of the required size has the same chance of being the selected sample.

5.45   **a.** The 366 possible dates (including February 29) from January 1 to December 31.
b. Answers may vary but an example is that they could write each date on a piece of paper, put them in a large paper bag, mix well, and select one from the bag.

5.47   **a.** The whole numbers from 1 to 49.
b. Answers will vary.
c. It doesn't matter because every set of six numbers has the same chance of occurring in each drawing.

5.49   **a.-c.** Answers will vary, and the method used will vary for different instructors. Most samples will give a mean and a median somewhere between 19 cm. and 21 cm.
Note: With Minitab, a random sample of values from a column can be selected using Calc>Random Data> Sample from columns.
d. It's possible, but all ten measurements would have to be 20.0 cm for the standard deviation to be 0 and this is unlikely. (The chance is approximately 1 in 67 million that all ten measurements are 20.0.)

5.51   Simple random sample.

5.53   **a.** Stratified sample: use the three types of schools as strata. Create a list of all students for each of the three strata. Draw a simple random sample from each of the three lists.
b. Cluster sample: Use individual schools or individual classes as clusters. Select a random sample of clusters, measure all (or a sample of) students in those clusters.
c. Simple random sample: Obtain a list of all students in the classes at all schools; take a simple random sample from that combined list.

5.55   Random digit dialing is more like a cluster sample because a sample of exchanges is found and then only numbers within those exchanges are sampled. It would be more like a stratified sample if numbers were sampled within every possible exchange.

5.57   **a.** All local taxpayers.
b. Parents of schoolchildren in the local schools.
c. By only including the parents of schoolchildren, local taxpayers who do not have children cannot be included in the sample. Their opinions about increasing taxes to support schools may differ from those of parents of school-age children.

5.59   **a.** Convenience sample.
b. Self-selected sample.

5.61   You would probably start with your friends. After that, you would probably ask people who seem friendly and approachable, rather than people who seem mean or in a rush. You may feel most comfortable asking people close to your own age, or people you perceive to be compatible with you in some way (perhaps even in their opinions). The polls failed to predict the winner because quota sampling is not likely to provide a sample representative of the population of interest.

5.63   **a.** The sampling frame is the collection of 66,000 dentists who subscribe to one or both of the two dental magazines. The subscription lists of these two magazines do not include all dentists in the population, and this could create a bias if the subscribers to the magazines are different than non-subscribers when it comes to using Bristol Myers products. This could be the situation, for instance, if Bristol Myers frequently advertises in the magazines.
b. Nonparticipation (nonresponse) is a serious problem here because only 1,983 dentists out of 10,000 selected responded.

c. They could have sent a reminder to those dentists who had not responded, reminding them to fill out the survey, or could have sent them another copy of the survey form. They could have phoned the dentists if their phone numbers were also available.

5.65   **a.** This is an example of a self-selected or volunteer sample. Magazine readers voluntarily responded to the survey, and were not randomly selected.
b. These results may not represent the opinions of all readers of the magazine. The people who respond probably do so because they feel stronger about the issues (for example, violence on television or physical discipline) than the readers who do not respond. So, they may be likely to have a generally different point of view than those who do not respond.

5.67   **a.** This is a self-selected (or volunteer) sample.
b. Probably higher, because people who would say they have seen a ghost would be more likely to call the late-night radio talk show than others. They might even be more likely to be listening to such a show.

5.69   Answers will vary. One possible restatement is, "Should former drug dealers be allowed to work in hospitals after they are released from prison, or not?"

5.71   Anonymous testing. The test results are not linked to the person's name.

5.73   Confidentiality and anonymity, and desire to please.

5.75   This will differ for each student. Any 2 questions in which one changes the way respondents would think about the other can be used. An example: "Are you aware that over 30% of homeless people in this city are mothers with children?" and "Do you think more public money should be used to help homeless people?"

5.77   **a.** Bias due to a desire to please the interviewer would most likely occur with a door-to-door interview because the respondent is in the presence of the interviewer.
b. Volunteer response is most likely to occur with a mail survey because people could easily throw the survey away if they don't want to fill it out.
c. Traditionally, it is thought that bias due to a perceived lack of confidentiality would most likely occur with a door-to-door interview and would least likely occur with a mail survey. However, many students who have answered this question thought that a perceived lack of confidentiality would most likely occur with a telephone interview. They said that with a phone interview, you could not judge whether to trust the interviewer because you can't see them, while with a door-to-door interview you might better be able to judge whether to trust the interviewer.

5.79   **a.** Closed-form.
b. No, because the researchers are assuming these are the movies most people say they like. Another movie, such as "The Wizard of Oz," could have drawn more votes than the movies given as choices.
c. Answers may vary. For open-form questions, one advantage is that the people can respond without being influenced by the movies listed. One disadvantage is that they may not remember the name of the movie they are thinking of and will have to put something else down. For closed-form questions, one advantage is that researchers can narrow down the choices if they are going to represent this data in a bar chart or pie chart. One disadvantage is like that mentioned in part b.

5.81   **a.** Open-form.
b. Yes, because his name was prominent in the media in that time period due to the 25th anniversary of his death.
c. Answers may vary, but probably lower considering the timing of the poll. By listing other stars by name, people would be less likely to focus on Elvis's name, which had been prominent in the media.

5.83   **a.** Your chance is $10/100 = .10$ (or 10%).
b. The answer will vary, but will be between 0 and 5 for about 99% of all students. You might expect to be in about 10% of 20 = 2 samples.

5.85   **a.** The answer will vary, It is likely to be in the vicinity of 68 inches.

       b. The answer will vary. It is likely to be about 4 or 5 inches (from about 66 to 70).

       c. The answer will vary.

       d. The idea is to compare the set of 20 sample means to 68 inches, the known population mean. Student answers will vary depending upon how "lucky" they were in their sampling and their standards for accuracy. Roughly, about 50% of the sample means will be further than 0.8 inches from 68, and about 20% will be further than 1.5 inches from 68 (the population mean). So, a sample of 10 observations may not be reliably accurate.

5.87   Answers may vary. Any ordering of these 10 two digit numbers will suffice: 00, 07, 15, 19, 24, 33, 44, 51, 65, and 99. These are the numbers corresponding to the numbered locations of the sample selected. One possible answer is 00071 51924 33445 16599.

5.89   The answer may vary, but in theory the mean of the 20 sample means is more likely to be close to the population mean (68) than most of the means for samples of 10 individuals. It is very likely that the mean of the 20 means will be within 0.5 inches of 68, one way or the other. Only about 30% or so of the individual sample means will be this close.

5.91   **a.** All adults in the U.S. at the time the poll was taken.

       b. $\dfrac{1}{\sqrt{1048}} = .031$ or 3.1%

       c. 34% ± 3.1%, or 30.9% to 37.1%.

5.93   These answers will differ somewhat for each student.

       a. Self-selected sample: Put an ad in the local paper asking people to fill out the survey.

       b. Deliberate bias: Send a questionnaire with wording such as "Don't you agree that there is too much trash in our streets and that more public trash containers are needed?"

       c. Ordering of questions: An initial question might be "Do you think there is too much trash on the streets of our city?" This could be followed by the question "Do you think there should be more trash containers available?"

       d. Desire to please: Send interviewers out for a door-to-door survey. With someone asking them face-to-face if they want more trash containers, participants may feel the need to please the interviewer and say "yes."

5.95   The "difficulty and disaster" is the use of a self-selected sample in the TV Guide poll.

5.97   **a.** .022 or 2.2%.

       b. Proportion: .17 ± .022 or .148 to .192; Percentage: 17% ± 2.2% or 14.8% to 19.2%.

5.99   **a.** Solving the equation $\dfrac{1}{\sqrt{n}} = .03$ gives n=1,111. Sample sizes close to this number result in a margin of error rounded off to 3%.

       b. *Sample percent ± Margin of error* is 55% ± 3%, or 52% to 58%.

       c. Answers will vary. An example is "In a survey of about 1000 adults, 55% of those asked favor gun control. From this result, we can conclude that it is likely that somewhere between 52% and 58% of the population of adults favor gun control.

5.101   **a.** The population of interest is the population of all students at the university.

       b. The sampling frame is the collection of all students enrolled in statistics classes at the time of the survey.

       c. The sample is the 500 students to whom the survey was mailed.

       d. The extent to which the sample represents the population of interest depends on what types of students are enrolled in statistics classes that term. Depending on the university requirements and the term in question, there may be more freshman or seniors than in the general student body, and certain majors are likely to be over-represented or under-represented.

5.103 To determine if the increase observed in the samples represents a real increase in the population, sample size or margin of error must be provided for each survey. The 20% versus 25% may be within the margin of error for the surveys.

5.105 The "Quickie Poll" would probably be most representative and fastest.

5.107 **a.** The margin of error is $\dfrac{1}{\sqrt{n}} = \dfrac{1}{\sqrt{1400}} = .027$, or 2.7%.

b. The interval of values that probably covers the true percentage of the population that saw the UFO is $20\% \pm 2.7\% = 17.3\%$ to 22.7%. The calculation is *sample percent ± margin of error*.

5.109 Nonparticipation bias was the most problematic in that survey. Only 34% of the scientists responded and the ones who did were typically white males over the age of 50. Other scientists who did not respond may feel differently about the topic.

5.111 **a.** Selection bias: Send the survey only to coffee bar owners.
b. Nonresponse bias: Send the survey to a legitimate random sample, but make the questions so outrageous that only those who support the position would respond; others would not take it seriously.
c. Response bias: Word the question to elicit the desired response. For example, "If coffee drinking were to be restricted to coffee bars, they would be more likely to provide attractive seating areas both inside and outside for coffee lovers to enjoy their coffee. Would you support such a plan?"

5.113 Answers will vary, but many websites have such polls.

5.115 No, the opinions of the 642 voters considered likely to vote are probably more representative of the population of voters who will participate in the November election.

5.117 For GOP voters within the sample, the margin of error would be larger because the sample size is smaller than for the entire sample.

5.119 **a.** The margin of error for arts and humanities is $\dfrac{1}{\sqrt{n}} = \dfrac{1}{\sqrt{166}} = .078$, or 7.8%.

The confidence interval is *Sample percentage ± Margin of error*, which is $66\% \pm 7.8\%$ or 58.2% to 73.8%. Expressed as an interval for a population proportion, the answer is $.66 \pm .078$ or .582 to .738.

b. The margin of error for engineering and sciences is $\dfrac{1}{\sqrt{n}} = \dfrac{1}{\sqrt{229}} = .066$, or 6.6%.

The confidence interval is *Sample percentage ± Margin of error*, which is $38\% \pm 6.6\%$ or 31.4% to 44.6%. Expressed as an interval for a population proportion, the answer is $.38 \pm .066$ or .314 to .446.
c. The observed percentages differ by a substantial amount, and even when taking margin of error into account, it is clear that a difference between the disciplines exists in the population. The upper endpoint of the confidence interval for the engineering and sciences population is 31.4%, which is substantially lower then the lower endpoint of the confidence interval for the arts and humanities population, which is 58.2%. The interval for the population percentage of arts and humanities faculty in favor reaches to almost 74%.

CHAPTER 6
ODD-NUMBERED SOLUTIONS

6.1   **a.** Observational study. Female students would not be assigned to be in a sorority or not.
     b. Randomized experiment. The doctor could randomly assign medications, and an experiment will provide stronger evidence about any difference in the effectiveness of the two medications.
     c. Randomized experiment. A random order for a server either introducing themselves or not could be used.
     d. Observational study. The psychologist could not assign specific amounts of television watching to the children.

6.3   **a.** Explanatory variable = sorority membership (or not); response variable = grade point average.
     a. Explanatory variable = medication used; response variable = extent of allergy relief for patient.
     a. Explanatory variable = server introducing self or not; response variable = tip amount (or percentage).
     a. Explanatory variable = television watching amount; response variable = bullying frequency.

6.5   **a.** Math skills and shoe size will both increase as children get older. The explanatory variable (shoe size) is related to age, and age also affects the response variable, math skills.
     b. Both the explanatory variable and the response variable may be related to time spent socializing. More social people may be more likely to get a cold because they sleep less and have many contacts with other people. More social people may also procrastinate in doing assignments.

6.7   **a.** Randomized experiment, because the experimenter randomly allocated students to type of course.
     b. Observational study, because participants choose whether they smoked or not.
     c. Observational study, because pre-existing groups are compared.

6.9   **a.** Unit = college student. Variables are procrastination habits and illness frequency.
     b. Unit = an SUV. Variables are manufacturer and damage sustained in crash test.

6.11   **a.** A randomized experiment is probably not possible here. Although it would be possible to randomly assign some people to long-term meditation, it would be hard to be sure that participants complied. Also, some assigned to the "no meditation" group may really want to meditate and would do so anyway. It would be much more practical to do an observational study in this situation.
     b. Yes, a randomized experiment could be done. People could be randomized to two groups—one that takes the special training program and one that does not. Afterward, the standardized test scores of the two groups could be compared.
     c. In part (a), the explanatory variable is whether or not a person participated in long-term meditation and the response variable is blood pressure. In part (b), the explanatory variable is whether or not a person took a special training program and the response variable is their score on a standard college admissions test.
     d. This will differ for each student. Some possible answers:
In part (a), one possible confounding variable is diet. People who practice long-term meditation may be more likely to follow a vegetarian diet, and this may affect blood pressure. In part (b), one possible confounding variable is intelligence. The people who sign up for the special training program may be the brightest students who are eager to get into prestigious universities.

6.13   **a.** This is a randomized experiment because children were randomly assigned to receive xylitol in syrup, gum, or lozenge form, or a placebo in gum or syrup form.
     b. The explanatory variable is what treatment was received (xylitol in the different forms or placebo in the different forms) and the response variable is whether or not a child got an ear infection.
     c. No, confounding variables are not likely to be a problem. Because children were randomly assigned to the different treatments, any child with a predisposition to ear infections was just as likely to be in the xylitol group as in the placebo group.

6.15   An observational study was done instead of an experiment because the researchers could not assign individuals either to attend a religious service once a week and pray regularly or to not engage in these practices.

6.17   Randomly divide the sixty participants into two groups of thirty. Assign one group to diet and the other to exercise.

6.19   **a.** Yes, because the sugar tablet is designed to look and taste like vitamin C.
             b. Yes, because the participants do not know which treatment they have been assigned.
             c. No, because the researchers know the treatment assignment for each participant.
             d. No, participants were not paired or matched in any way.
             e. No, participants are measured only once (at the end of the two month period).
             f. Yes, 10 students were assigned to each group.

6.21   **a.** Randomly divide stores into two groups of 10 and then assign a different method to each store.
             b. Pair the stores based on weekly sales or store size. In each pair, randomly assign one store to each method.

6.23   **a.** Form pairs of participants by age. Members of a pair would have similar ages. Randomly assign each member of a pair to use a different method.
             b. A matched-pair design is a good idea because we have the problem of memorization decreasing with age and we have such an age variation.

6.25   Answers may vary. You could number volunteers 1 to 100. Then use the software to randomly permute the integers from 1 to 100. Assign volunteers with the first 25 numbers in the permuted list to the first treatment, volunteers with the next 25 numbers in the permuted list to the second treatment, and so on.

6.27   **a.** This experiment was single-blind because the technicians who read the meters did not know the rate plan for a particular customer. The customers did know which rate plan they had. It was on their bill.
             b. This was a double-blind experiment because the participants did not know which tea they were drinking, and the psychologists did not know who was drinking the herbal tea.
             c. This experiment was neither single-blind nor double-blind. Everyone knew which of the three packaging designs was being used in each store. It could be physically seen.

6.29   **a.** The explanatory variable is whether or not special rates were offered and the response variable is electricity use during peak hours.
             b. The explanatory variable is whether or not the herbal ingredient was taken and the response variable is the change mood or level of depression after one month of drinking the tea.
             c. The explanatory variable is the packaging design used and the response variable is how much was sold in two months.

6.31   **a.** A control group was used. The researchers measured electrical use of 100 customers who did not receive the special rates.
             b. A control group and a placebo treatment were used. The control group was the 50 participants who drank the regular tea without the herbal ingredient. A placebo treatment was used because although the control group did not take the herb, they did drink tea every day.
             c. Neither a control group nor a placebo was used.

6.33   **a.** The teacher might record attendance information throughout the term.
             b. At the end of the term, ask students to remember and report their attendance for the term.
             c. The disadvantage of the prospective study is that student behavior might be altered if they know that the teacher is keeping attendance records. The disadvantage of the retrospective study is that students may not accurately remember or report their attendance habits.

6.35   **a.** Yes, because we are looking into the future of the different jobs economics majors had 10 years after graduation.
b. No, because the participants are not being asked to recall any past events.
c. No, because there is no control group (for instance, students without this degree).

6.37   Gather a sample of people over 50 who have skin cancer (cases) and a sample of people over 50 who do not have skin cancer (controls). Ask each individual to recall his or her lifetime sun exposure.

6.39   Although in general a randomized experiment is desirable it would be nearly impossible in this case. We would have to do a observational study as students will have made their own choices about where to live.

6.41   **a.** This was an observational study because 50 couples who owned cats or dogs were compared with 50 couples who did not. Couples were not assigned to own a pet or not. It might be considered to be a case-control study, with pet owners as cases and non-owners as controls. However, typically a case-control study is one for which the case/control distinction is for the response variable and in this study pet ownership was the explanatory variable.
b. The explanatory variable was whether or not the couple owned a pet and the response variables were marriage satisfaction and stress levels.
c. This will differ for each student. The careers of the couple could be confounding variables. People who work long hours and are already very stressed from work may not have enough time to take care of a pet. Their stressful careers may also make them less likely to have a satisfying marriage.
d.

Figure for Exercise 6.41d

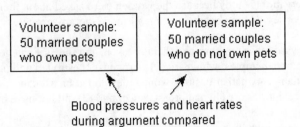

6.43   A randomized experiment provides stronger evidence of a cause-and-effect relationship.

6.45   **a.** This is an observational study because vegetarians and non-vegetarians are compared and these groups occur naturally. People were not assigned to treatment groups.
b. Since this is an observational study and not a randomized experiment, we cannot conclude that a vegetarian diet causes lower death rates from heart attacks and cancer. Other variables not accounted for may be causing this reduction.
c. This answer will differ for each student. One potential confounding variable is amount of exercise. This is a confounding variable because it may be that vegetarians also exercise more on average and this led to lower death rates from heart attacks and cancer.

6.47   Interacting. Smoking status affects the amount of difference in the risks of high blood pressures for women taking oral contraceptives versus those who do not

6.49   The Hawthorne effect.

6.51   The elephants are in captivity so their behavior will differ from what is in the wild.

6.53   **a.** Yes. The participants may not have good memories of lifelong exposure, and may be inclined to create memories of exposure if they have lung problems.

b. Yes. Because this is an observational study, confounding is always a concern. However, the most obvious alternative cause for lung problems is controlled because none of the participants have ever smoked. So, this source of confounding might not be a problem.

c. No. The participants are not in an experiment so the problem so this won't be a problem.

6.55   This will differ for each student. Here are some examples.
a. The *Hawthorne effect* may be a problem because the customers were treated in a special way.
b. *Ecological validity* may be a problem. In real world use, people may not drink a cup of the tea every day.
c. The *Hawthorne effect* could be a problem because the participating grocers might promote the product more enthusiastically than usual (because they want to please the manufacturer)
Extending results inappropriately is a possible problem. If all cities were large, then the results may not extend to people in smaller cities and towns.

6.57   **c.** *Relying on memory* is likely to be a problem. It would be hard for people to accurately remember the fat content of their childhood diet. *Confounding variables* also are likely to be a problem. For instance, people who have high fat diets might exercise less than those with low fat diets. If so, it would be difficult to separate the effects on heart disease of diet and exercise.

6.59   The term "effect modifier" makes sense because whether or not there were other smokers in the home modified the effect of the nicotine patch. When there were no other smokers in the home, 58% of the nicotine patch users were able to quit. When there were other smokers in the home, only 31% of the nicotine patch users were able to quit.

6.61   This was an observational study because the women were asked about their use of hormone therapy, so the researchers only observed the already existing hormone therapy groups.

6.63   The explanatory variable is the hormone therapy taken, and this is a categorical variable. The categories are the possible hormone therapies, which are (1) estrogen, (2) estrogen and progestin, and (3) no hormones. The response variable is whether or not a woman developed breast cancer. This is also a categorical variable and the categories are (1) had breast cancer and (2) did not have breast cancer.

6.65   This is an example of interacting variables. A woman's weight alters the effect of the hormone therapy. Although in the overall sample, hormone therapy was associated with an increased risk of breast cancer, this relationship was not found in heavier women.

6.67   This is an example of extending results inappropriately. If women today take lower levels of hormones than the women in the study, these results may not apply to women currently on hormone therapy.

6.69   This was a randomized experiment because participants were randomly assigned to one of three treatments—they were told to think about when they had bad hair, or told to think about bad product packaging, or not told to think about anything.

6.71   The study could not have been double blind because the participants had knowledge of their assigned task. The study could have been blind. It would have been if the researcher(s) who gave the self-esteem and self-judgment tests did not know the group assignments. Not enough information was given, however, to determine whether the study was blind or not.

6.73   This is an example of interacting variables. The effect on self-esteem of thinking about their bad hair is different for men than for women.

6.75   In some situations it may be impossible and/or unethical to carry out a randomized experiment. As an example, suppose that we wanted to study the connection between body weight and blood pressure. We could not assign individuals to gain a certain amount of weight so that we could see how it affected their blood pressure.

6.77   This is a completely randomized design because each child was randomly assigned to a different gym group (treatment). There were no natural matches of children or treatments and no blocks were established for the children, such as age groups.

6.79   **a.** Yes, a placebo and a double-blind procedure can be present in the same study. An example is the testing of a new medication given in pill form. A placebo pill that looks like the (new) experimental pill could be used. If neither the patient nor the evaluator of the patient know which pill the patient is taking, the study is double-blind.
b. Yes. In a case-control study, which is basically a retrospective observational study, pairing can be done by matching each case with a similar control.
c. No, this is not possible. A case control study is a type of observational study, so by definition, treatments or conditions are not assigned by the researchers Random assignment of treatment is done in randomized experiments.

6.81   **a.** The individual unit is a tomato plant. The two variables measured were the number of tomatoes produced and whether the tomato plant was raised in full sunlight or partial shade.
b. The individual unit is an automobile. The two variables measured were the gas mileage and whether the tires were under-inflated or inflated to their maximum possible pressure.
c. The individual unit is a classroom. The two variables measured were the number of children who did better than average on standardized tests and whether or not the classroom took a morning fruit snack break.

6.83   This is an observational study so a figure similar to Figure 6.2 is appropriate.

Figure for Exercise 6.83

6.85   Yes, a variable can be both a confounding variable and an interacting variable. As an example, suppose that the incidence of lung disease is compared for people who worked around asbestos versus people who did not work around asbestos and that asbestos workers were more likely to smoke than the other workers. This would make smoking a confounding variable. Suppose also that the effect of smoking versus not smoking on lung disease was much greater for asbestos workers than the other workers. This would mean that smoking is also an interacting variable.

6.87   **a.** The treatments were (1) telling the experimenters had maze bright rats, and (2) telling the experimenters they had maze dull rats.
b. The 12 individual experimenters, because we are assigning a treatment (telling them the rats are maze bright or maze dull) to the experimenters.
c. Answers may vary. One possibility: Give each experimenter some rats that are claimed to be maze bright and also some that are claimed to be maze dull.

49

6.89   **a.** Yes, a variable can be both a confounding variable and a lurking variable. By definition, a lurking variable is a variable that is not measured and accounted for in the interpretation of a study. It may be a confounding variable, an interacting variable, or some other variable that is related to the results.
b. No, a variable cannot be both a response variable and a confounding variable. A confounding variable affects the response variable and its effect cannot be separated from the effect of the explanatory variable.
c. No, a variable cannot be both an explanatory variable and a dependent variable. A dependent variable is thought to depend on the explanatory variable.

6.91   Confounding variables are more of a problem in observational studies than in randomized experiments. When participants are randomly assigned to treatments groups, they are just as likely to be in one treatment group as another. Any potential confounding variables are likely to be balanced over the treatment groups. For example, if age affects the response variable, randomization makes it probable that all the treatment groups will have both young and old participants. If an observational study is done and age is related to the explanatory variable, there may be generally older participants in one group and generally younger participants in another. The effects due to age and the effects due to the explanatory variable then cannot be easily separated.

6.93   **a.** Each student was measured twice.
b. The order of swearing or not swearing should be randomized.
c. There were $n = 64$ participants.

CHAPTER 7
ODD-NUMBERED SOLUTIONS

7.1   Random Circumstance: Flight arrival time for a randomly selected flight on one of the top 18 U.S. airlines in January 2010. The probability that the flight arrives on time (or early) is .787.

7.3   1/16 = .0625. After four students have been selected, sixteen remain as candidates, each with an equal chance to be picked.

7.5   Random Circumstance 1: Song on the radio when first turned on
   ➢ Robin's favorite song is playing
   ➢ Robin's favorite song is not playing
   Random Circumstance 2: Color of traffic light when Robin approaches the main intersection
   ➢ Traffic light is green when Robin arrives
   ➢ Traffic light is red or yellow when Robin arrives
   Random Circumstance 3: Nearest available parking space
   ➢ Robin finds an empty parking space in front of the building
   ➢ Robin does not find an empty parking space in front of the building

7.7   **a.** Answers will vary, but one example is whether a ticket for a lottery is going to be a winner, for a lottery that will occur in the future.
   b. Answers will vary, but one example is whether a ticket for a "scratcher" lottery game is a winner; in these games the player scratches off a covering to reveal whether the ticket is a winner or not.

7.9   1000/125000 =1/125, or .008

7.11   **a.** Relative frequency probability; based on "observing relative frequencies of outcomes over many repetitions of the same situation."
   b. Personal probability.
   c. Relative frequency probability; based on "measuring a representative sample and observing relative frequencies of possible outcomes."

7.13   *For the first circumstance*, Robin could repeatedly note whether or not her favorite song is playing when she first turns on the radio. The probability that her favorite song is playing is the number of times her favorite song is playing divided by the total number of days she did this.
   For the second circumstance, Robin could repeatedly note whether or not the traffic light was green when she arrived, and divide the number of times it was green by the total number of days she did this.
   For the third circumstance, Robin could repeatedly note whether or not she quickly found an empty parking spot in front of the building, and divide the number of times she found a good parking spot by the total number of times she did this.

7.15   An individual could determine his or her probability of winning this game by playing it a large number of times and recording how many games he or she won out of the total number of games played.

7.17   No, this does not mean that Alicia will be called on to answer the first question exactly once during the semester. In the long run, if this statistics class had many meetings, the proportion of times that Alicia would be called on for the first question is 1/50. This semester she may be called on 0, 1, 2, or even more times during the 50 class meetings.

7.19   Of the children who slept in darkness, the number with myopia or high myopia is 15 + 2 = 17.
   So, the probability = 17/172 = .0988 that a randomly selected child who slept in darkness would develop some degree of myopia.

7.21   The simple events are the possible number of days, {0, 1, 2, 3, 4, 5, 6, 7}

7.23   **a.** Yes, because they don't contain any of the same outcomes (simple events). Part of the definition of A^c is that it does not contain any of the same simple events as A.
b. No, they are dependent events. If A occurs then A^C cannot occur. If A does not occur, then A^C must occur. For independent events, knowing that one has occurred does not give us any information about the probability that the other one occurs. For events that are complements, knowing that A occurred definitely gives us information about $P(A^C)$ because we know that $P(A^C) = 0$.

7.25   **a.** Yes. The probability for each event is between 0 and 1, and over all possible outcomes, the sum of the probabilities equals 1.
b. No. The sum of the probabilities exceeds 1.
c. Yes. The sum of the probabilities is less than 1, and each individual probability is between 0 and 1.

7.27   **a.** Yes, B and C are independent because the outcome of the drawing one week in unrelated to the outcome the next week.
b. Now B and C are not independent. If Vanessa wins in week 1, her card from that week will no longer be part of the drawing the next week, thus changing her probability of winning in week 2.

7.29   **a.** Mutually exclusive (red die cannot be both 3 and 6), but not independent.
b. Independent, but not mutually exclusive. Knowing the outcome of one of the dice does not provide any information about the probabilities for the other one, so they are independent. The red die can be 3 in the same toss that green die is 6, so they are not mutually exclusive.

7.31   On any given day, 3 of the 50 students are chosen to answer questions. If the drawing is fair, for any one of the three questions each of the 50 students has the same probability of being chosen as any other student. So, prior to any draws the probability is 3/50 that Alicia will be one of the 3 students chosen.

7.33   **a.** P(A and B) = 0. The number 5 is not even, so events A and B cannot both happen.
b. No. Once we know that one of the events has occurred, the probability that the other one occurred is 0.
c. Yes, A and B are mutually exclusive. The same student cannot pick both 5 and an even number.

7.35   Age and fertility status are not independent. The probability of being fertile changes with age.

7.37   **a.** For each of C_1, C_2, and C_3, the unconditional probability is 1/50. Prior to any selections, each of the 50 students has the same chance to be picked for any question.
b. $P(C_3|C_1) = 0$. If Alicia is picked for the first question, her name is taken out of the bag so she cannot possibly be picked for the third question.
c. No, C_1 and C_3 are not independent events. They are dependent because $P(C_3) = 1/50$, but $P(C_3|C_1)=0$. The probability for C_3 changes if it is known that C_1 has happened.

7.39   **a.** A^C = not getting all 3 heads in the 3 tosses of the coin. Another way to state this is that A^C is getting 0, 1 or 2 heads in the 3 tosses. Another way is that it is the event that at least one tail occurs in the three tosses.
b. 7/8. This can be found as 1− P(A) = 1−(1/8) = 7/8.

7.41   **a.** {Monday, Tuesday, Wednesday, Thursday, Friday, Saturday, Sunday}
b. 2/7, using Rule 2b.

7.43   **a.** A and B are mutually exclusive because the first digit cannot be both even and odd.
b. A and C, or B and C. These are independent because knowing whether the first digit is even or odd does not change the probability that the second digit is odd.
c. Yes, A and B are complements. The first digit must be either even or odd.

7.45   **a.** No, they are not independent. P(A in both classes) ≠ P(A in English)×P(A in history), as it would for independent events.
b. P(A in either English or history) = P(A in English) + P(A in history) − P(A in both classes) = .70 + .60 − .50 = .80.

7.47 **a.** With replacement, because each digit can be 0, 1, ..., 9 even if that number has been used.
b. Without replacement, because 3 different students need to be chosen.
c. Without replacement, because 5 different people are needed.

7.49 **a.** Probability = 11/12 that the first stranger does not share your birth month.
b. Probability = 11/12 that the second stranger does not share your birth month.
c. Probability = (11/12)(11/12) = 121/144 = .84 that neither shares your birth month. Use the multiplication rule for two independent events (Rule 3b).
d. P(at least one) = 1–P(neither) = 1–.84 = .16, or 23/144. The event that at least one of the two shares your birth month is the complement of the event that neither does.

7.51 P(one of each) = 1– P(two of same kind) = 1 – .5002 = .4998. The event that she has one of each sex is the complement of the event that she has two children of the same sex.

7.53 **a.** P(both are friends of president) = (10/40)(10/40) = 1/16 = .0625.
The probability of picking a friend is the same for each draw because sampling is with replacement.
b. P(both are friends of president) = (10/40)(9/39) = .0577.
Note that the second probability (9/39) is the *conditional* probability of a friend *given* that the first selection was a friend.
c. P(neither are friends of president) = (30/40)(30/40) = .5626
The probability of not picking a friend is the same for each draw because sampling is with replacement.
d. P(neither are friends of president) = (30/40)(29/39) = .5577
Note that the second probability (29/39) is the *conditional* probability the second selection is not a friend *given* the first selection is not a friend.

7.55 **a.** The numbers are as follows:

Magazine Type	International	National only	Total
News	20	10	30
Sports	5	15	20
Total	25	25	50

b. P(includes international news | news magazine) = 20/30 = .67.
c. 25/50 = .50. There are 50 magazines; 20 include international news and 5 include international sports.

7.57 **a.** The tree diagram is below.

Figure for Exercise 7.57a

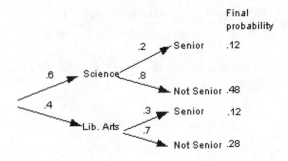

b. The desired probability is the sum of the "final probabilities" for the two paths that end in "Senior" which is .12 + .12 = .24. Therefore 24% of the students are seniors.

7.59 **a.** P(A) = .80; P(A and B) = .25.
b. P(B|A) = P(A and B)/P(A) = .25/.80 = .3125.
c. P(B^c|A) = 1 – P(B|A) = 1 – .3125 = .6875.

7.61   Define events for the first draw as A = first card is an ace, A^c = first card is not an ace.
Define events for the second draw as B = second card is an ace, B^c = second card is not an ace.
Then P(first card is an ace and second card is a non-ace) = P(A)P(B^c| A) = (1/13)(48/51) = .0724.
The relevant probabilities were found as P(A) = 4/52 = 1/13, and P(B^c| A) = 48/51 because once we know
that the first draw is an ace, there are 51 cards remaining, of which 48 are not aces. You could also use a
tree diagram to solve this problem.

7.63   Based on the information in the table in the previous exercise, the probability is 15/40 = .375. A tree
diagram could also be used. Make the first set of branches correspond to the two different classes (junior,
senior) and the second set of branches correspond to taking calculus or not.

Figure for Exercise 7.63

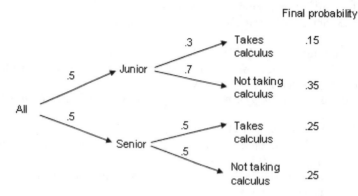

The desired conditional probability is P(junior | taking calculus).

$$P(\text{junior} | \text{taking calculus}) = \frac{P(\text{junior and taking calculus})}{P(\text{taking calculus})}.$$ This is Rule 4 for conditional probability.

A student taking calculus is either a junior or a senior, so (using Rule 2b):
$P(\text{taking calculus}) = P(\text{junior and taking calculus}) + P(\text{senior and taking calculus})$

Each element of the previous formula can be found using the multiplication rule (Rule 3a).
$P(\text{junior and taking calculus}) = P(\text{junior}) \times P(\text{taking calculus} | \text{junior}) = (.5)(.3) = .15$
$P(\text{senior and taking calculus}) = P(\text{senior}) \times P(\text{taking calculus} | \text{senior}) = (.5)(.5) = .25$

This leads to $P(\text{taking calculus}) = .15 + .25 = .40$.

So, $P(\text{junior} | \text{taking calculus}) = \dfrac{P(\text{junior and taking calculus})}{P(\text{taking calculus})} = \dfrac{.15}{.40} = .375$

7.65   The tree diagram illustrates the desired probability as well as probabilities for all other outcomes. Figure for
this exercise is on the next page.

Figure for Exercise 7.65

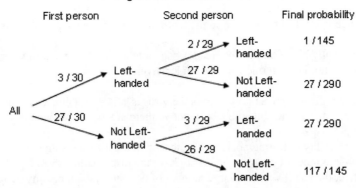

7.67 Number of occurrences / number of simulations = 45/10000 = .0045.

7.69 **a.** Estimated probability = 8/50 = .16. There were 8 simulations in which prize 4 was received 3 or more times.
b. Estimated probability = 7/31 = .2258. There were 31 simulations in which all four prizes were not received. Prize 4 was received 3 or more times in 7 of these 31 simulations.
c. Estimated probability is 1/50 = .02. There was only one simulation in which the full collection of four prizes was received and prize 4 was received at least 3 times.

7.71 For probability of a correct guess = .20 instead of .30, change the simulation procedure so that only two digits (8 and 9 for example) are counted as a correct guess while the other eight digits are counted as a wrong guess.

7.73 **a.** Probability = 73/1000 = .073 that 3 digits chosen at random will sum to 15. Of the 1000 possible three-digit numbers, 73 have digits that sum to 15. To determine this answer, list (or imagine listing) the three-digit numbers for which the digits sum to 15. Do so in a patterned way. Suppose the first digit is 0. If so, the next two digits must sum to 15, and there are four ways for this to occur (6+9, 7+8, 8+7, 9+6). Suppose the first digit is 1. If so, the next two digits must sum to 14, and there are five ways for this to occur (5+9, 6+8, 7+7, 8+6, 9+5). Continue in this manner considering each possible digit as the first digit. The following table summarizes the method just described. The values in the third column sum to 73.

1st digit	Sum of 2nd and 3rd digits	Number of ways
0	15	4
1	14	5
2	13	6
3	12	7
4	11	8
5	10	9
6	9	10
7	8	9
8	7	8
9	6	7

b. This will differ for each student.
c. This answer is a conjecture so anything that sounds sensible is correct. Generally, a collection of people given the task of creating numbers randomly do not create a batch of random numbers. So, it's likely that the proportion picking three digits that sum to 15 will be different from the theoretical probability for doing so. In this case, it might be that the proportion would be higher than the probabilities found in parts (a) and

55

(b). In an attempt to pick "randomly," many people might tend to balance out their digits so the average is an "average" possible digit (which is about 5). If the three digits have average=5, the sum would be 15.

7.75   No. Each toss is independent of the others, so the tenth toss will still have a probability of .5 of being heads.

7.77   All sequences are equally likely because P(Heads) = P(Tails) = .5. The probability for each sequence is $(1/2)^5 = 1/32$ by the multiplication rule for independent events.

7.79   **a.** The probability that the person actually carries the virus is 1/11 = .0909 because in the low-risk population, for every infected person who tests positive there are 10 people who test positive and do not carry the virus.
b. Although the probability of testing positive for those with the disease is high, the reverse is not true. If there are a very large number of people who do not have the disease, then even if only a small percentage of those test positive, the result will be a large number of positive tests in healthy people. The table below conveys this idea. Notice that everyone with the disease tests positive, and almost 90% of those without the disease correctly test negative, yet of every 11 people who test positive, only one has the virus.

	Test positive	Test negative	Total
HIV	10	0	10
No HIV	100	890	990
Total	110	890	1000

7.81   **a.** Probability = 1/365 = .0027 that your teammate's mother would have the same birthday as you.
b. The easiest way to find this is to first find the probability that none of the 5 family members has the same birthday as you. Then, subtract that probability from 1 (because "none" and "at least one" are complementary events).The probability that none of the five family members has the same birthday as you is $\left(\dfrac{364}{365}\right)^5 = .9864$. (Use the multiplication rule along with the fact that the probability is 364/365 that a single member does not match.) So the desired probability is 1 − .9864 = .0136.
c. No, it is not. While that specific event has low probability, it is quite possible that someone in the class would have a birthday matching a teammate's family member's birthday. With 30 class members, each with 5 family members, probability = $1-(.9864)^{30} \approx .34$ that someone in the class shares a birthday with a teammate's family member.

7.83   No, she is not correct. Assuming birth outcomes are independent, having 3 consecutive boys does not change the probability that the fourth child will be a boy.

7.85   This will differ for each student.

7.87 P(match on at least one topic) = $1- P$(match on none of the topics) = $1-\left(\dfrac{49}{50}\right)^{10} = .1829$.

Notice that the probability of no matches is calculated using the multiplication rule for independent events (Rule 3b extension). It would not be a total surprise to match on at least one topic because at least one match will occur in about 18% of the occasions in which two strangers compare details for 10 topics.

7.89   Simulation relies on the relative frequency interpretation. Because personal probability requires one to give a personal assessment of the likelihood of an outcome, it is not something that can be simulated by a computer.

7.91   **a.** For males, P(*either* Rationalist *or* Teacher) = P(Rationalist) + P(Teacher) = .15 + .12 = .27.
b. For females, P(not Teacher) = 1−.14 = .86.
c. For males, P(both roommates are Rationalists) = (.15)(.15) = .0225.
d. For females, P(both roommates are Rationalists) = (.08)(.08) = .0064.

56

e. If two roommates are both Rationalists, they could *either* both be males *or* both be females. So P(both Rationalists) = P(Males and Rationalists) + P(Females and Rationalists).

P(Males and Rationalists) = P(Males)$\times P$(Rationalists | Males) = (.5)(.0225) = .01125

P(Females and Rationalists) = P(Females) $\times P$(Rationalists | Females) = (.5)(.0064) = .0032

So, P(both Rationalists) = .01125 + .0032 = .01445.

7.93   **a.** P(no failure) = $1-P$(failure) = $1- (1/10000)$ = $9999/10000$ = .9999.

b. P(no failures among 4 plugs) = $(.9999)^4 = .9996$. This is an application of the multiplication rule for independent events. We are finding the probability that the first *and* second *and* third *and* fourth spark plugs do not fail.

c. P(at least one fails) = $1 - P$(no failures) = $1 - .9996$ = .0004.

7.95   **a.** Percentage = 40%. Written as a probability it is P(get A | regular attendance) = .40.

b. Percentage = 10%. Written as a probability it is P(get A | not regular attendance) = .10.

c. Students who get an A *either* regularly attend *or* do not. Over the whole class:

P(get A) = P(regularly attend *and* get A) + P(do not regularly attend *and* get A)

The two components on the right side of the previous equation are:

P(regularly attend and get A) = P(regularly attend) $\times P$(get A | regularly attend) = (.7)(.4) = .28

P(do not regularly attend and get A) = P(do not regularly attend) $\times P$(get A | do not regularly attend) = (.3)(.1) = .03

So, P(get A) = .28 + .03 = .31, which is 31%.

7.97   The table is as follows.

Attendance	A	Not A	Total
Regular	28,000	42,000	70,000
Not regular	3,000	27,000	30,000
Total	31,000	69,000	100,000

7.99   **a.** Probability = 0.5 that a randomly selected person is above the median. Recall that the median of a data set divides the ordered data into two equal halves. (The problem stated that none of the values were exactly equal to the median.)

b. P(all 4 above median) = $\left(\dfrac{1}{2}\right)^4 = \dfrac{1}{16} = .0625$. The multiplication rule for independent events (Rule 3b extension) is used because we want to know the probability that the first *and* second *and* third *and* fourth students all are above the median.

c. No, because the events are not independent. For example, the probability that the second person selected has a value above the median depends upon whether the first person selected had a value above the median or not.

7.101   Most likely, this figure is based on a combination of personal probability, long run relative frequency and physical assumptions about the world. We are told that it is based on "integration of scientific fact and expert opinion." Data about how often powerful earthquakes have occurred in California in the past was probably taken into account, along with physical assumptions about earthquake faults, and expert knowledge of earthquakes.

7.103   **a.** P(both are) = P(first is)$\times P$(second is) = (.35)(.35) = .1225.

This assumes independence, and the multiplication rule (Rule 3b) is used because we want the probability that the first *and* second person are both naturalized citizens.

b. If a married couple (both foreign-born) was selected, the citizenship status of the husband and wife almost certainly would not be independent. If one of these individuals decided to become a U.S. citizen, it most likely would affect the probability that the other decides this as well.

c. The probabilities reported by the Census Bureau are relative frequency probability. They observed a large number of foreign-born people in the U.S. in 1997 and noted what proportion of those were naturalized citizens.

57

d. The statement as written refers to the same person but uses the probabilities as if referring to two independent people. A correct version would be: "If a foreign-born person in the U.S. was randomly selected in 1970, the probability he or she was a naturalized citizen would have been .64. By 1997, if a foreign-born person in the U.S. was randomly selected the probability he or she would have been a naturalized citizen had dropped to .35.

7.105   Approximate probability = 485/1669 = .2906 that a randomly selected person smokes.

7.106   Approximate probability = 612/1669 = .3667 that a randomly selected person has been divorced.

7.107   Given the person has been divorced, approximate probability = 238/612 = .3889 that he or she smokes. This estimate is the proportion that smokes within the Divorced =Yes column of the table.

Note for exercises 7.109-7.111: The questions ask about randomly selected members of the entire population of adult Americans who have ever been married, not just the 1669 in the sample. So, the probabilities for problems 7.109 to 7.111 can be found as if the two people were selected with replacement. For such a large population, the divorce status of the first randomly selected person will not have any noticeable affect on the probability the second person is divorced. However, these were not meant to be trick questions, and answers based on using only the sample are also provided. For those answers, we assume that sampling is *without* replacement.

7.109   There are two possible (mutually exclusive) sequences in which one person smokes and the other does not.

Sequence 1: first smokes, second does not;   probability $= \left(\dfrac{485}{1669}\right)\left(\dfrac{1184}{1669}\right)$

Sequence 2: first does not smoke, second does ;   probability $= \left(\dfrac{1184}{1669}\right)\left(\dfrac{485}{1669}\right)$

The two possibilities have equal probability, so answer $= 2 \times \left(\dfrac{485}{1669}\right)\left(\dfrac{1184}{1669}\right) = .4123$.

Based on sample only the probability is .4125

7.111   Using Rule 4 for unconditional probability, $P(\text{both} \mid \text{at least one}) = \dfrac{P(\text{both and at least one})}{P(\text{at least one})}$.

$P(\text{both and at least one}) = .1345$ (found in Exercise 7.110) because "both and at least one" is the same as "both."

$P(\text{at least one is divorced}) = 1 - P(\text{neither is divorced}) = 1 - \left(\dfrac{1057}{1669}\right)\left(\dfrac{1057}{1669}\right) = 1 - .4011 = .5989$.

So, the answer is $P(\text{both} \mid \text{at least one}) = \dfrac{.1345}{.5989} = .2246$.

Some students may misinterpret the question to be "Given that the first person had been divorced, what is the probability that the second one had also been divorced?" and argue that it is .3667, the same as the unconditional probability that the second person had been divorced, because the divorce status of the two people are independent.

CHAPTER 8
ODD-NUMBERED SOLUTIONS

8.1   **a.** Discrete
        b. Continuous
        c. Discrete
        d. Discrete

8.3   **a.** Discrete
        b. Continuous
        c. Continuous
        d. Discrete

8.5   This will differ for each student. An example of another continuous random variable is the amount of rainfall (in inches) during the event time. Rainfall can be any number between 0 and some maximum value. One example of another discrete random variable is the number of police, ambulance, and fire vehicles driving by with sirens. The number of vehicles can be 0, 1, 2,…, and so on, up to the logical maximum for the situation.

8.7   For discrete random variables, the probability that the variable equals a specified value can be found. For continuous random variables, we can only determine the probability that the value of the variable falls into a specified interval.

8.9   **a.** Answer = 3/6 = ½. This makes the probabilities add to 1.
        b. Answer =.1. This makes the probabilities add to 1. Calculation is $1 - (.1 + .2 + .3 + .3) = .1$.

8.11   **a.**

Meals, X	1	2	3	4
Cumulative Prob.	.10	.10+.32 = .42	.10+.32+.56 = .98	1

        b. .42, the cumulative probability for 2 meals.

8.13

k	0	1	2	3
$P(X=k)$	1/8	3/8	3/8	1/8

8.15   **a.** $.06 + .13 = .19$
        b. $P(X > 0) = 1 - P(X = 0) = 1 - .73 = .27$.
        Another method is to add the probabilities for X = 1, 2, 3, 4.
        c.

k	0	1	2	3	4
$P(X \leq k)$	.73	.89	.95	.98	1

8.17   The answer is 18/36=1/2. Add probabilities given in Example 8.10 for X = 2, 4, 6, 8, 10, and 12.

8.19   **a.** <u>Condition 1:</u> Sum = $.1 + .1 + .3 + .5 = 1$.
        <u>Condition 2:</u> Each of the four probabilities is between 0 and 1.

b.

Figure for Exercise 8.19c

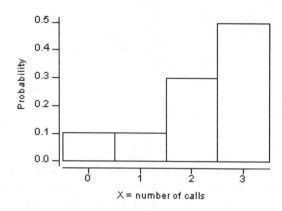

c.

k	0	1	2	3
P(X≤k)	.1	.2	.5	1

As an example, $P(X \leq 1) = P(X=0) + P(X=1) = .1 + .1 = .2$.

8.21   The probability that $X=1$ is the probability that the first child is a girl, so $P(X=1)=.5$. For $X=2$, the sequence must be Boy, Girl so $P(X=2)= (.5)(.5)=.25$. For $X=3$, the sequence is Boy, Boy, Girl and this probability is $(.5)(.5)(.5)=.125$. The probability that $X=4$ can be found be subtracting the sum of the other probabilities from 1, and this gives $P(X=4)=.125$. A summary of the distribution is:

a. Simple events are {G, BG, BBG, BBBG, BBBB}

b. Probability of G = .5, probability of BG = .5×.5 = .25, probability of BBG = .5×.5×.5 = .125, probability of BBBG = 5×.5×.5 ×.5= .0625, and probability of BBBB = .0625

c. For X = number of children, the probability distribution is:

k	1	2	3	4
P(X=k)	.5	.25	.125	.125

The probability for X = 4 is the sum of the probabilities for BBBG and BBBB.

d.

Figure for Exercise 8.21d

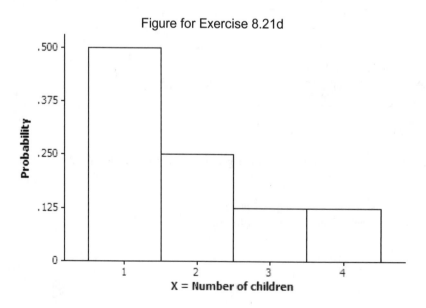

8.23   This may differ for each student and the explanation should support the response. Following are some suggestions.
a. The *cdf* would be of more interest, since she would want to know if the number (and percent) of constituents who oppose the new law is less than or equal to some standard (like 50%).
b. The *pdf* may be of more interest because it reveals what proportion of people become infected on the first exposure, what proportion of people become infected on the second exposure, etc. The *cdf* may be of interest to someone who has had *k* exposures and wants to know the probability of infection so far.

8.25   **a.**

X = amount won/lost	+$4.00	−$2.00
Probability	.3	.7

b. $E(X) = (+\$4.00)(.3) + (-\$2.00)(.7) = -\$0.20$
c. In the long run (over infinitely many plays) the player will lose an average of 20 cents per game.

8.27 $E(X) = (\$100)(.01) + (-\$5)(.99) = -\$3.95$.

8.29   **a.** $E(X) = (0)(.512) + (1)(.384) + 2(.096) + 3(.008) = 0.6$
b. Over (infinitely) many repeats of playing 3 times, the mean number of wins (per 3 games) is 0.6.

8.31   **a.** $E(X) = -2(.25) + 0(.50) + 2(.25) = 0$.
b. $\sigma^2 = (-2 - 0)^2(.25) + (0 - 0)^2(.50) + (2 - 0)^2(.25) = 2$.
c. $\sigma = \sqrt{2} = 1.414$

8.33   An intuitive solution (and a correct one) is that the expected value for the number of girls among three children is $3 \times .5 = 1.5$. A more elaborate solution is to determine the probability distribution for X = number of girls, and then use the formula for E(X). The distribution given in Example 8.6 is:

k	0	1	2	3
$P(X=k)$	1/8=.125	3/8 =.375	3/8=.375	1/8=.125

The expected value is calculated as the sum of "value×probability" and the calculations are, $E(X) = (0 \times .125) + (1 \times .375) + (2 \times .375) + (3 \times .125) = 1.5$ girls.

8.35 $\sigma = \$29.67$. The standard deviation is calculated as $\sigma = \sqrt{\sum(x - \mu)^2 p(x)}$ where the sum is over all values of *x*. The expected value (mean) was found in Example 8.12 to be $\mu = -\$0.35$. The details for finding σ are:

x	$4,999	$49	$4	$0	−$1
$x - \mu$	$4,999.35	$49.35	$4.35	$0.35	−$0.65
$p(x)$	.000035	.00168	.0303	.242	.726

$$\sqrt{4999.35^2(.000035) + \$49.35^2(.00168) + \$4.35^2(.0303) + \$0.35^2(.242) + \$0.65^2(.726)}$$
$$= \$29.67$$

8.37 $\mu = E(X) = \sum xp(x) = (500 \times .1) + (0 \times .9) = \$50$ per person (in the long run).

8.39   **a.** $E(X) = 0(.04) + 1(.26) + 2(.70) = 1.66$
Note that the probability for 1 parent is the sum of the probabilities for mother only and father only
b. No, in the sense that no child could be living with 1.66 parents.

8.41   **a.** Yes. $n = 200$ and $p = .5$.
b. $E(X) = np = (200)(.5) = 100$.

8.43   **a.** $n = 30$ and $p = 1/6$.
b. $n = 10$ and $p = 1/100$.
c. $n = 20$ and $p = 3/10$.

8.45   The answers can be found using any of the methods discussed in Section 8.4, including the use of Minitab or Excel.
a. $P(X = 5) = .0264$.
b. $P(X = 2) = .3020$.
c. $P(X = 1) = .2684$.
d. $P(X = 9) = .000004$.

8.47   The formulas are $\mu = np$ and $\sigma = \sqrt{np(1-p)}$
a. $\mu = 10(1/2) = 5$ and $\sigma = \sqrt{10(.5)(1-.5)} = 1.5811$
b. $\mu = 100(1/4) = 25$ and $\sigma = \sqrt{100(.25)(1-.25)} = 4.33$
c. $\mu = 2500(1/5) = 500$ and $\sigma = \sqrt{2500(.2)(1-.2)} = 20$
d. $\mu = 1(1/10) = .1$ and $\sigma = \sqrt{1(.1)(1-.1)} = .3$
e. $\mu = 30(.4) = 12$ and $\sigma = \sqrt{30(.4)(1-.4)} = 2.683$

8.49   **a.** The probability of success does not remain the same from one trial (game) to the next. The probability of winning a game against a good team is not the same as the probability of winning a game against a poor team.
b. The number of trials is not specified in advance.
c. The probability of success does not remain the same from one trial to the next because whether or not the first card is an ace affects the probability that the next card is an ace, and so on. This also means that trials are not independent.

8.51   **a.** .32805; the "by hand" calculation is $\dfrac{5!}{1!\,4!}(.1)^1(.9)^4$; in Excel, BINOMDIST(1,5,.1,0)
b. $P(X \le 1) = .91854$; in Excel, BINOMDIST(1,5,.1,1)
c. $P(X \ge 2) = 1 - P(X \le 1) = 1 - .91854 = .08146$

8.53   The answers for this exercise can be found using any of the methods discussed in Section 8.4, including the use of Minitab or Excel.
a. $P(X = 4) = .2051$
b. $P(X \ge 4) = 1 - P(X \le 3) = 1 - .6496 = .3504$

8.55   **a.** Note that 1/4 of 1,000 is 250 so the desired probability is $P(X \ge 250)$. $n = 1000$ and $p =$ the proportion of adults in the United States living with a partner, but not married at the time of the sampling. The value of p is not known.
b. The desired probability is $P(X \ge 110)$, $n = 500$, and $p = .20$.
c. Note that 70% of 20 is 14 so the desired probability is $P(X \ge 14)$. $n = 20$, and $p = .50$.

8.57   **a.** .3 (because the interval 0 to 3 is 3/8 of the possible outcomes and the distribution is uniform)
b. .4 (because the interval 4 to 8 is 4/10 of the possible outcomes and the distribution is uniform)
c. .5 (because the interval 5 to 10 is 5/10 of the possible outcomes and the distribution is uniform)

8.59   **a.** X is a uniform random variable (and it is continuous).
b. X ranges from 0 to 100 and the area under any density curve is 1, so $f(x) = 1/100 = .01$ for all x between 0 and 100. This creates a rectangle (with area=1) similar to Figure 8.2.
Note: $f(x) = 0$ for any x not between 0 and 100.

62

c. $P(X \leq 15$ seconds) is the area of the rectangle from 0 to 15 seconds. The interval width is 15 and the height is 1/100, so the answer is $(15)(1/100) = .15$.

d. $P(X \geq 40$ seconds) is the area of the rectangle between 40 and 100. The interval width is 60 and the height is 1/100 so the answer is $(60)(1/100) = .60$.

e.

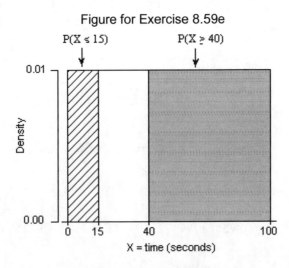

Figure for Exercise 8.59e

f. The expected value or mean is 50. The distribution is symmetric, so the mean equals the median. For a uniform random variable, the median is at the middle of the range of possible values.

8.61 This will differ for each student. The answer must be something that is equally likely to fall anywhere in an interval. An example is the position of the second-hand on a clock when you glance at it.

8.63
a. $\dfrac{1.5 - 0}{1} = 1.5$.

b. $\dfrac{4 - 10}{6} = -1$.

c. $\dfrac{0 - 10}{5} = -2$.

d. $\dfrac{-25 - (-10)}{15} = -1$.

8.65 Table A.1 can be used to find the answers.
a. .5000
b. .3632
c. .6368

8.67
a. Answer = .8413. For 200 lbs, $z = \dfrac{200 - 180}{20} = 1$. $P(Z \leq 1) = .8413$.

b. Answer = .2266. For 165 lbs, $z = \dfrac{165 - 180}{20} = -0.75$. $P(Z \leq -0.75) = .2266$.

c. Answer = .7734. This is the "opposite" event to part (b), so calculation is $1 - .2266 = .7734$.

8.69
a. $P(Z \leq -1.4) = .0808$
b. $P(Z \leq 1.4) = .9192$
c. $P(-1.4 \leq Z \leq 1.4) = P(Z \leq 1.4) - P(Z \leq -1.4) = .9192 - .0808 = .8384$
d. $P(Z \geq 1.4) = 1 - P(Z \leq 1.4) = 1 - .9192 = .0808$. Equivalently, $P(Z \geq 1.4) = P(Z \leq -1.4) = .0808$.

63

8.71   **a.** $z = \dfrac{71-75}{8} = -0.5$. So, $P(X \le 71) = P(Z \le -0.5) = .3085$.

b. $z = \dfrac{85-75}{8} = 1.25$. So, $P(X \ge 85) = P(Z \ge 1.25) = 1 - P(Z < 1.25) = 1 - .8984 = .1016$.

Equivalently, $P(Z > 1.25) = P(Z < -1.25) = .1016$.

c. For pulse $= 59$, $z = \dfrac{59-75}{8} = -2$ while for pulse $= 95$, $z = \dfrac{95-75}{8} = 2.5$. Find the area under the standard normal curve between these two z-scores.
$P(-2 \le Z \le 2.5) = P(Z \le 2.5) - P(Z \le -2) = .9938 - .0228 = .9710$.

8.73   First, find the standardized score z^* for which $P(Z \le z^*) = .10$. It's $z^* = -1.28$ so the answer is 1.28 standard deviations below the mean. The answer is $(-1.28 \times 8) + 75 = 64.76$, or about 65.

8.75   **a.** For 10, $z = \dfrac{10-18}{6} = -1.33$ so $P(X < 10) = P(Z < -1.33) = .0918$

b. For 30, $z = \dfrac{30-18}{6} = 2$ so $P(X > 30) = P(Z > 2) = 1 - P(Z \le 2) = 1 - .9772 = .0228$. Equivalently, $P(Z > 2) = P(Z < -2) = .0228$.

c. For 21, $z = \dfrac{21-18}{6} = 0.5$ while for 15, $z = -0.5$. So, $P(15 \le X \le 21) = P(-0.5 \le Z \le 0.5) = P(Z \le 0.5) - P(Z \le -0.5) = .6915 - .3085 = .3830$

d. $P(X > 35) = P(Z > \dfrac{35-18}{6}) = P(Z > 2.83) = 1 - P(Z \le 2.83) = 1 - .9977 = .0023$.

Equivalently, $P(Z > 2.83) = P(Z < -2.83) = .0023$.

8.77   **a.** $z^* = -1.96$. If using Table A.1, look for .025 within the interior part of the table.

b. $z^* = 1.96$. If using Table A.1, look for .975 within the interior part of the table. Or, note that the area to the right of z^* must be .025, so by the symmetry of the standard normal curve the answer is the positive version of the answer for part (a).

c. $z^* = 1.96$ because if .95 is in the central area, .975 must be the area to the left of z^*. This means the answer is the same as for part (b).

8.79   The value of z^* for which $P(Z \le z^*) = .25$ is about -0.675. Look for .25 in the interior of Table A.1. The height with this z-score is $(-0.675 \times 2.7) + 65 = 63.2$ inches. The percentile ranking for a height of 63.2 inches is .25 or 25%.

8.81   **a.** Answer $= .0571$.
For a binomial random variable with $n = 50$ and $p = .512$,
$\mu = np = 50(.512) = 25.6$, and $\sigma = \sqrt{50(.512)(1-.512)} = 3.535$.
Thus, for $X = 20$, $z = \dfrac{20-25.6}{3.535} = -1.58$.
$P(X \le 20) \approx P(Z \le -1.58) = .0571$.

b. Answer $= .0749$. With the continuity correction, we find $P(X \le 20.5)$.
For $X = 20.5$, $z = \dfrac{20.5-25.6}{3.535} = -1.44$. So, $P(X \le 20.5) = P(Z \le -1.44) = .0749$.

8.83   **a.** Answer $= .9015$. For a binomial random variable with $n = 1000$ and $p = .60$,

64

$\mu = np = 1000(.60) = 600$, and $\sigma = \sqrt{1000(.60)(1-.60)} = 15.492$.

For $X = 620$, $z = \dfrac{620-600}{15.492} = 1.29$. $P(Z \le 1.29) = .9015$.

8.85 **a.** Mean $= np = 200(.7) = 140$; standard deviation $= \sqrt{np(1-p)} = \sqrt{200(.7)(1-.7)} = 6.481$

b. .0322 without continuity correction, or .0384 with continuity correction. Without continuity correction, z = (128 – 140)/6.481 = –1.85 and in Table A.1 the associated cumulative probability is .0322. With continuity correction, z = (128.5 –140)/6.481 = –1.77 and in Table A.1 the associated cumulative probability is .0384. If you use a TI calculator or software, your answer will differ slightly in the fourth decimal place from the one given here.

8.87 **a.** The mean for the total of four individuals is the total of the means for individuals. This is $100 + 100 + 100 + 100 = 400$ seconds.
b. The variance for an individual is $(10)^2 = 100$, so the variance for the total of four independent individuals $= 100 + 100 + 100 + 100 = 400$. The standard deviation for the total is $\sqrt{400} = 20$ seconds.
c. For $T = 360$, $z = \dfrac{360-400}{20} = -2$. Thus, $P(T < 360) = P(Z < -2) = .0228$.

8.89 **a.** $1/4 = .25$
b. $n = 7$; $p = .25$
c. $n = 8$; $p = .25$
d. $n = 7 + 8 = 15$; $p = .25$

8.91 **a.** For X, the mean is $\mu_X = np = 10 \times .5 = 5$ and the variance is $\sigma_X^2 = np(1-p) = 10 \times .5 \times .5 = 2.5$.
For Y, the mean is $\mu_Y = np = 20 \times .4 = 8$ and the variance is $\sigma_Y^2 = np(1-p) = 20 \times .4 \times .6 = 4.8$.
For $X+Y$, the mean is $\mu_X + \mu_Y = 5 + 8 = 13$, the variance is $\sigma_X^2 + \sigma_Y^2 = 2.5 + 4.8 = 7.3$ and the standard deviation is $\sigma = \sqrt{7.3} = 2.702$. The name of the distribution of the sum cannot be given because p is different for X and Y.
b. For $X+Y$, the mean is $100+50 = 150$, the variance is $15^2 + 10^2 = 325$, and the standard deviation is $\sqrt{325} = 18.028$. The distribution of the sum is normal because X and Y both have normal distributions.
c. For Y, the mean is $np = 200 \times .25 = 50$ and the variance is $np(1-p) = 200 \times .25 \times .75 = 37.5$. For X+Y, the mean is $100+50 = 150$, the variance is $15^2 + 37.5 = 262.5$, and the standard deviation is $\sqrt{262.5} = 16.202$. The name of the exact distribution cannot be given (although it is approximately normal because X is normal and Y is approximately normal due to its values of n and p).

8.93 **a.** For $n=10$, the mean (expected number) is $np = 10 \times .25 = 2.5$. For $n=20$, the mean is $np = 20 \times .25 = 5$, and for $n=50$, the mean is $np = 50 \times .25 = 12.5$.
b. For 80 trials, the expected number correct is 20 if Joe is just guessing. This can be determined in two different ways. It is the sum of the means (2.5, 5, and 12.5) for the three separate experiments. And, it also is the mean for a single binomial experiment with $n=80$ and $p= .25$. The combination of three experiments can be viewed as a single binomial experiment because p, the chance of success on a trial, is the same in all experiments.
c. For both $n=10$ and $n=20$, the binomial distribution should be used because $np < 10$ in both cases. For $n=50$, the normal curve can be used to approximate the answer although either Minitab or Excel could be used to determine an exact answer based on the binomial distribution. The exact answers based on binomial distributions are:
$n = 10$ experiment, $P(X \ge 4) = 1 - P(X \le 3) = 1 - .7759 = .2241$
$n = 20$ experiment, $P(X \ge 8) = 1 - P(X \le 7) = 1 - .8982 = .1018$

65

$n = 50$ experiment, $P(X \geq 20) = 1 - P(X \leq 19) = 1 - .9861 = .0139$

Note: To find the approximate solution for the n=50 experiment, use a normal approximation. For a binomial experiment with $n = 50$ and $p = .25$, the mean is 12.5 and the standard deviation is

$\sqrt{50 \times .25 \times (1-.25)} = 3.06$. The z-score for 20 correct is $z = \dfrac{20-12.5}{3.06} = 2.45$.

So, $P(X \geq 20) = P(Z \geq 2.45) = 1 - P(Z < 2.45) = 1 - .9929 = .0071$.

d. In a binomial distribution with $n = 80$ and $p = .25$, $P(X \geq 32) = 1 - P(X \leq 31) = 1 - .9978 = .0022$. Or, use a normal approximation to find an approximate answer.

For a binomial experiment with $n=80$ and $p=.25$, the mean is $np = 20$ and the standard deviation is

$\sqrt{80 \times .25 \times .75} = 3.87$. The z-score for 32 correct is $\dfrac{32-20}{3.87} = 3.10$.

So, $P(X \geq 32) = P(Z \geq 3.10) = 1 - P(Z \leq 3.10) = 1 - .9990 = .0010$.

e. It would be better for Joe if he showed the combined data. In part (d) we see that the probability is only .002 (2 in a thousand) that a person who was just guessing could get 32 or more correct in 80 trials. This is more impressive evidence than the results for the three separate experiments.

8.95 Let X = train time and Y = bus time. If the train is faster, $X < Y$ which is equivalent to $X - Y < 0$. For $X - Y$ (the difference in times) the mean is 60 min. $-$ 50 min. $=$ 10 min. and the standard deviation is

$\sqrt{\sigma_X^2 + \sigma_Y^2} = \sqrt{2^2 + 8^2} = 8.246$ min. For 0 min., $z = \dfrac{0-10}{8.246} = -1.21$. From Table A.1, $P(Z < -1.21) = .1131$.

8.97 The histogram will probably not have a bell-shape because the different means for men and women may create a bimodal distribution (one with two peaks). The bimodal effect will not be as clear, however, as intuition might suggest. Taller than average women and shorter than average men combine to create frequent observations of heights between the two means. Here's what the theoretical population distribution looks like:

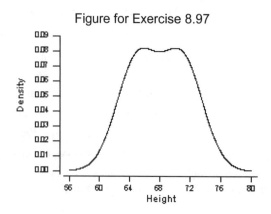

Figure for Exercise 8.97

8.99 **a.**

k	$70	$-$2
$P(X=k)$	1/38	37/38

b. Calculate the sum of "value×probability" over all values.

$E(X) = (\$70)(1/38) + (-\$2)(37/38) = -\$4/38 = -\$0.1053$. In the long run, the average outcome will be that players lose about 10.5 cents to the casino per \$2 bet.

8.101 $E(X) = (3 \times .07) + (4 \times .14) + (5 \times .52) + (6 \times .25) + (7 \times .02) = 5.01$ courses per student

66

8.103   **a.** The probability of success is not the same for each trial.

b. The trials are not independent because the chance of selecting a woman on a particular selection is affected by the results of previous selections. This also means the chance of a "success" is not the same on all trials.

8.105   **a.** Using Minitab, Excel, or other software that can determine probabilities for a binomial distribution with $n=200$ and $p=.6$, the exact answer is $P(X \geq 140) = 1 - P(X \leq 139) = 1 - .9979 = .0021$.

An approximation using the normal distribution is as follows: The mean and standard deviation are

$$\mu = np = 200(.6) = 120 \text{ and } \sigma = \sqrt{np(1-p)} = \sqrt{200(.6)(1-.6)} = 6.928 \text{ . For } 140, \ z = \frac{140-120}{6.928} = 2.89 \text{ so the}$$

approximate answer is $P(Z \geq 2.89) = 1 - P(Z < 2.89) = 1 - .9981 = .0019$.

b. Note that 70% of 20 is 14 so the desired probability is $P(X \geq 14)$. The exact answer using a binomial distribution with $n= 20$ and $p=0.6$ is $P(X \geq 14) = 1 - P(X \leq 13) = 1 - .75 = .25$.

An approximation using the normal distribution is as follows: The mean and standard deviation are

$$\mu = np = 20(.6) = 12 \text{ and } \sigma = \sqrt{np(1-p)} = \sqrt{20(.6)(1-.6)} = 2.191 \text{ . For } 14, \ z = \frac{14-12}{2.191} = 0.91 \text{ so the}$$

approximate answer is $P(Z \geq 0.91) = 1 - P(Z < 0.91) = 1 - .8186 = .1814$.

8.107   **a.** The distribution of the number of girls is a binomial distribution with $n=4$ and $p=0.5$. The probabilities, which can be found using any of the methods described in Section 8.4, are:

k = number of girls	0	1	2	3	4
$P(X=k)$	.0625	.25	.375	.25	.0625

b. The distribution for Karen will be the same as it is for Kim.

c. The distribution for T, the total number of girls, is a binomial distribution with $n=8$ and $p=0.5$. The 8 children the two sisters plan to have are 8 independent trials for which the chance of a "success" (a girl) is 0.5 on each trial. This answer can also be justified by recognizing that T is the sum of two independent binomial random variables (X and Y) with the same p.

8.109   **a.** .6915; $z = (98.6 - 98.2)/0.8 = 0.5$, in Table A.1 the associated cumulative probability is .6915.

b. .1056; $P(X > 99.2) = 1 - P(X \leq 99.2)$, for 99.2, $z = (99.2 - 98.2)/0.8 = 1.25$, so we find $1 - P(z \leq 1.25) = 1 - .8944 = .1056$

c. .3345; $P(97 \leq X \leq 98) = P(X \leq 98) - P(X \leq 97) = P(z \leq -0.25) - P(z \leq -1.5) = .4013 - .0668 = .3345$

8.111   If 1000 repetitions are done, the simulated answer is likely to be in the range .06 to .10. The theoretical probability is .079. **Minitab Tip:** Use **Calc>Random Data>Binomial** to generate 1000 rows in C1 for a binomial with $n=10$ and $p=.8$. Use that command again, to generate 1000 rows in C2 for n=10 and p=.2. Use **Calc>Calculator** to create the sum C3=C1+C2. Finally, use **Stat>Tables>Tally** to get the distribution of the total score in C3.

8.113   The desired probability is $P(X>75)$ where X = vehicle speed.. For 75 mph, $z = \frac{75-67}{6}$ so $P(X>75) =$

$P(Z>1.33) = 1 - P(Z<1.33) = 1 - .9082 = .0918$.
Equivalently, $P(Z>1.33) = P(Z<-1.33) = .0918$.

CHAPTER 9
ODD-NUMBERED SOLUTIONS

9.1   **a.** Statistic because it is a sample value.
b. Parameter because it is a population value.
c. Statistic because it is a sample value
d. Parameter because it is a population value (errors in the entire manuscript).

9.3   The truth refers to the population parameter. The sample statistic is computed from the data, so we know its value. It is the population parameter we are trying to estimate.

9.5   **a.** The parameter of interest is the proportion of the population that has as least one copy of the E4 allele for Apo E.
b. A confidence interval would make more sense. There is no obvious null value because there is no "chance" value of interest to test.
c. The scientists could not possibly have measured everyone in the population. They are relying on a sample to estimate the percentage of the population that has the allele.

9.7   **a.** Parameters. Because everyone has been measured, the means are population values.
b. Yes, because the population means are known exactly, the shareholders can be certain that the mean salary for men in the company is $1500 higher than the mean salary for women.

9.9   **a.** $\hat{p}$ because this is a sample proportion.
b. p because this is a population proportion.
c. $\hat{p}$ because this is a sample proportion.

9.11   *Hint:* This is an example of Situation 1 on page 318 of the text.
a. Research question: What proportion of parents in the school district support the new program?
b. Population parameter: p = proportion of all parents in the school district who support the new program
c. Sample estimate: $\hat{p} = \dfrac{104}{300} = .347$

9.13   *Hint:* This is an example of Situation 5 on page 319 of the text.
a. How much difference is there between the mean number of CDs owned in the populations of male and female students?
b. Parameter = $\mu_1 - \mu_2$, where μ_1 = population mean number of CDs owned for males and μ_2 = population mean number of CDs owned for females.
c. $\bar{x}_1 - \bar{x}_2 = 110 - 90 = 20$.

9.15   Research question: How much difference is there between the proportions getting relief from sore throat symptoms in the population if herbal tea is used versus if throat lozenges are used?
Population parameter: $p_1 - p_2$ = difference in proportions reporting relief if everyone in the population were to use herbal tea compared with if everyone in the population were to use throat lozenges.
Sample estimate: $\hat{p}_1 - \hat{p}_2$ = difference in observed proportions reporting relief for the two different methods in the study of 200 volunteers.

9.17   Paired data. Two different variables are measured for each individual and interest is in the amount of difference.

9.19   Answers will vary. The response variable should be categorical, and the difference between the proportions having a particular trait or opinion in two independent groups should be the focus.
An example: The difference between the proportions of Democrats and Republicans who are in favor of a new tax law.

9.21   **a.** The population of interest is teens who go out on dates, and the parameter of interest is p = the proportion of that population who would say they had been out with someone of another race or ethnic group. The sample statistic is $\hat{p}$ = .57. (You could answer using percents instead.)

b. The population of interest is people who slept in darkness as babies and the parameter of interest is p = the proportion of that population who were free of myopia later in childhood. The sample statistic is $\hat{p}$ = .90.

c. The parameter of interest is $p_1 - p_2$ = the difference in proportions of myopia in later childhood for the populations who slept in darkness as babies and who slept in full light as babies, respectively. The sample statistic is $\hat{p}_1 - \hat{p}_2 = .90 - .45 = .45$.

d. The parameter of interest is μ = the mean height for the population of married British women in 1980. The sample statistic is $\overline{x}$ = 1602 millimeters.

9.23   The histogram of the 100 SAT scores would be more spread out. The spread of the sampling distribution gets smaller as the sample size increases. Think of a histogram of the original sample as equivalent to the sampling distribution for the mean for samples of size 1, which would have much larger spread than the sampling distribution for the mean for samples of size 100.

9.25   **a.** The mean of the sampling distribution of the sample proportion $\hat{p}$ is the population proportion p.

b. One value from the sampling distribution of $\hat{p}$ is one sample proportion, denoted by $\hat{p}$.

9.27   **a.** The population mean would not change.
b. The sample mean would change for each sample.
c. The standard deviation of $\overline{x}$ would not change. It is based only on population information and the sample size, which is 50 and does not change from one sample to the other.
d. The standard error of $\overline{x}$ would change because it is an estimate of the standard deviation of the sampling distribution and uses sample data, which changes for each sample.
e. The sampling distribution of $\overline{x}$ remains the same for all samples of the same size (50 in this case) from the same population.

9.29   **a.** The mean of the sampling distribution of the sample mean is μ, the population mean, and it remains the same.
b. The sampling distribution of the sample mean remains approximately normal.
c. The standard deviation of the sampling distribution depends on the population standard deviation and the sample size, both of which remain the same, so it remains the same as well. In fact all features of the sampling distribution remain the same, because it depends on the population and the sample size, not on the sample itself.

9.31   **a.** Not likely. As shown in Figure 9.1, the mean will almost surely be between 6.6 and 7.4 hours.
b. This would be possible. As shown in Figure 9.1, the mean could be as high as 7.4 hours, so there are some likely values in the interval 7 to 8 hours.
c. The distribution of original sleep hours is much more spread out than the sampling distribution shown in Figure 9.1, and an individual value between 8 and 9 hours would be quite likely. (Remember that the mean is 7.1 hours and the standard deviation is 2 hours.)
d. It is very unlikely that the maximum would be less than 8 hours. The sample of sleep hours had a mean of 7.1 hours and a standard deviation of 2 hours. The maximum is likely to be between 2 and 3 standard deviations above the mean.

9.33   **a.** Mean $= p = .5$; s.d. $\left(\hat{p}\right) = \sqrt{\dfrac{.5(1-.5)}{400}} = .025$

b. Mean $= p = .5$; s.d. $\left(\hat{p}\right) = \sqrt{\dfrac{.5(1-.5)}{1600}} = .0125$

70

9.35   A change in the sample size does not affect the value of the mean of the sampling distribution. Increasing the sample size decreases the value of the standard deviation. For a four-fold increase in sample size the standard deviation is cut in half.

9.37   **a.** Mean $= p = .55$.

b. s.d. $\left(\hat{p}\right) = \sqrt{\dfrac{.55(1-.55)}{100}} = .0497 \approx .05$.

c. About $.55 \pm (3)(.05)$, or about .40 to .70. This is the interval Mean $\pm$ 3 s.d., which is the part of the Empirical Rule that says that about 99.7% of the values will be within three standard deviations of the mean.

9.39   **a.** $\hat{p} = 300/500 = .60$.

b. $s.e.(\hat{p}) = \sqrt{\dfrac{\hat{p}(1-\hat{p})}{n}} = \sqrt{\dfrac{.60(1-.60)}{500}} = .022$.

9.41   **a.** Mean $= .70$; s.d. $\left(\hat{p}\right) = \sqrt{\dfrac{.70(1-.70)}{200}} = .0324$.

b. $.70 \pm (3 \times .0324)$, or .6028 to .7972.

c. $\hat{p} = 120/200 = .60$. This is a statistic.

d. The value .60 is slightly below the interval of possible sample proportions for 99.7% of all random samples of 200 from a population where $p = .70$. In other words, the sample proportion is unusually low if the true proportion were .70. Maybe the population value is actually less than .70.

9.43   The mean is .05 and the standard deviation is $s.d.(\hat{p}) = \sqrt{\dfrac{p(1-p)}{n}} = \sqrt{\dfrac{.05(1-.05)}{400}} = .011$. About 95% of all samples will have a sample proportion in the interval $.05 \pm (2)(.011)$, which is .028 to .072 (approximately .03 to .07). The sampling distribution of $\hat{p}$ is approximately a normal distribution, so about 95% of sample proportions will fall in the interval $p \pm 2\, s.d.(\hat{p})$.

9.45   **a.** $p = .20$ (20% expressed as a proportion).

b. $\hat{p} = 18/100 = .18$.

c. $s.e.(\hat{p}) = \sqrt{\dfrac{\hat{p}(1-\hat{p})}{n}} = \sqrt{\dfrac{.18(1-.18)}{100}} = .0384$. Notice that the standard error calculation uses the sample proportion $\hat{p}$.

d. The mean of the sampling distribution of $\hat{p}$ equals the population proportion, which is .20.

e. $s.d.(\hat{p}) = \sqrt{\dfrac{p(1-p)}{n}} = \sqrt{\dfrac{.20(1-.20)}{100}} = .04$.

9.47   The standard error of $\hat{p}$ is calculated using an observed value of a sample proportion and it (the standard error) *estimates* the "true" standard deviation of the sampling distribution of $\hat{p}$. The standard deviation is found using the known (or assumed to be known) value of the population proportion.

The formula for the standard error is $s.e.(\hat{p}) = \sqrt{\dfrac{\hat{p}(1-\hat{p})}{n}}$.

The formula for the standard deviation is $s.d.(\hat{p}) = \sqrt{\dfrac{p(1-p)}{n}}$.

In practice, the standard error will be used more often because the value of the population proportion usually will not be known.

9.49   The mean is $p_1 - p_2$ which is 0 if the two proportions are equal.

9.51   **a.** The standard deviation is $\sqrt{\dfrac{p_1(1-p_1)}{n_1} + \dfrac{p_2(1-p_2)}{n_2}} = \sqrt{\dfrac{.3(.7)}{500} + \dfrac{.36(.64)}{500}} = .0297$

    b. $\sqrt{\dfrac{.3(.7)}{100} + \dfrac{.2(.8)}{100}} = .0608$

    c. $\sqrt{\dfrac{.4(.6)}{800} + \dfrac{.4(.6)}{200}} = .0387$

9.53   **a.** Condition 1 is met because sample proportions will be available for independent random samples. Condition 2 will be met as long as the proportions who favor the candidate are between .01 and .99 for each population, thus ensuring that all of the quantities $n_1 p_1$, $n_1(1 - p_1)$, $n_2 p_2$, $n_2(1 - p_2)$ are at least 10.
    b. Condition 1 is not met; the researcher is not taking separate random samples from two populations. The two sample proportions described will not be independent.
    c. No, the sample sizes are not large enough. With only 10 observations for each method, it is impossible to have all of the quantities $n_1 p_1$, $n_1(1 - p_1)$, $n_2 p_2$, $n_2(1 - p_2)$ be at least 10.
    e. The appropriate region is to the left of 0.

9.55   **a.** The mean of the sampling distribution would not change because it does not depend on the sample sizes.
    b. The approximate shape (normal) would not change as long as the samples were large.
    c. The standard deviation of the sampling distribution would change because the sample sizes are part of the formula. (Larger samples result in a smaller standard deviation.)

9.57   **a.** The mean could be specified; it is $p_1 - p_2 = 0$.
    b. The approximate shape could be specified because appropriate conditions are met for it to be approximately normal.
    c. The standard deviation could be specified because the population proportions and the sample sizes are known.
    d. The standard error could not be specified because it is calculated using the sample proportions, which would not be known before the samples are taken.

9.59   **a.** Yes, this falls into Situation 1.
    b. No, for a skewed population, this sample size is too small.
    c. Yes, this falls into Situation 2.

9.61   **a.** $s.d.(\bar{x}) = \dfrac{\sigma}{\sqrt{n}} = \dfrac{24}{\sqrt{16}} = 6$.

    b. $s.d.(\bar{x}) = \dfrac{\sigma}{\sqrt{n}} = \dfrac{24}{\sqrt{64}} = 3$.

    c. Increasing the sample size decreases the value of the standard deviation of the sampling distribution of the sample mean. A fourfold increase in sample size cut the standard deviation in half.

9.63   **a.** $\bar{x}$ and s. These are sample statistics.

    b. $s.e.(\bar{x}) = \dfrac{s}{\sqrt{n}} = \dfrac{47.7}{\sqrt{28}} = 9.014$.

9.65   **a.** The distribution of individual weights for passenger + luggage is approximately normal with mean of 210 pounds and standard deviation of 25 pounds.

Figure for Exercise 9.65a

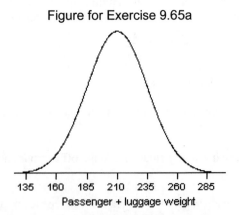

Passenger + luggage weight

b. The sampling distribution of possible sample means for random samples of n= 40 is approximately a normal distribution. The mean is $\mu = 210$ pounds.

The standard deviation of the sampling distribution is $s.d.(\bar{x}) = \dfrac{\sigma}{\sqrt{n}} = \dfrac{25}{\sqrt{40}} = 3.95$ pounds.

c. The sampling distribution of the mean has a much smaller standard deviation than the distribution of individual weights. Remember that about 99.7% of the distribution is within three standard deviations of the mean. For individual weights (including luggage) about 99.7% of the distribution is in the interval 210 ± 75 pounds. For possible sample means, about 99.7% of the distribution is in the interval 210 ± 11.85 pounds.

Figure for Exercise 9.65c

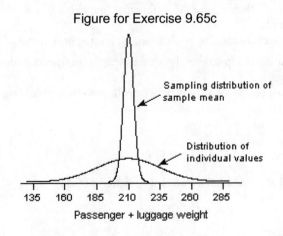

Sampling distribution of sample mean

Distribution of individual values

Passenger + luggage weight

d.   If the total weight is 8800, the mean weight is 8800/40 = 220 pounds. The question asked is equivalent to asking what is the probability that the *mean* weight is greater than 220 pounds for 40 passengers. Because the question is about a sample *mean*, use the sampling distribution described in part (b) to find the answer for $P(\bar{x} > 220)$.

For $\bar{x}$ =220 pounds, $z = \dfrac{220 - 210}{3.95} = 2.53$.

Use Table A.1 to find that $P(Z \le 2.53) = .9943$.

$P(\bar{x} > 220) = P(Z > 2.53) = 1 - P(Z \le 2.53) = 1 - .9943 = .0057$

Notice that this probability is 57 in 10,000, which is equivalent to 1 in 175 (divide 10,000 by 57). In the long run, about 1 of every 175 sold-out flights will exceed the total weight limit.

9.67   The standard error of the sample mean is calculated using the standard deviation of the measurements in an observed sample, and it (the standard error) *estimates* the "true" standard deviation of the sampling distribution of the sample mean. The standard deviation is found using the known (or assumed to be known) value of the standard deviation of the population of measurements.

The formula for the standard error is $s.e.(\bar{x}) = \dfrac{s}{\sqrt{n}}$.

The formula for the standard deviation is $s.d.(\bar{x}) = \dfrac{\sigma}{\sqrt{n}}$.

In practice, the standard error will be used more often because the value of the population standard deviation usually will not be known.

9.69   Two independent samples. The dormitory students and the off-campus students constitute two distinct, unrelated samples.

9.71   **a.** The parameter is the difference in population means for independent samples, $\mu_1 - \mu_2$. Males and females are independent groups, not paired in any way.
b. The parameter is the mean of paired differences, μ_d. Each female is measured at the beginning and end of basic training.

9.73   No. The mean is the population mean of the paired differences, μ_d, which will be 0 only if the means for the two measurements are the same.

9.75   **a.** The parameter of interest is the population mean of paired differences, μ_d. For each person in the population of men similar to the ones in this experiment the difference of interest is hypothetical – the difference in number of times he would drift in two hours of driving after consuming alcohol and after getting too little sleep.
b. The sample statistic is $\bar{d}$, the mean of the differences in number of lane shifts under the two conditions for the 12 men in the sample.
c. The population of differences must be bell-shaped (at least approximately).
d. The sampling distribution is approximately normal with mean 0 and standard deviation $5/\sqrt{12} = 1.44$.

9.77   No. Both populations must be approximately bell-shaped, or both sample sizes must be large.

9.79   **a.** The mean is $\mu_1 - \mu_2 = 8 - 4.5 = 3.5$ days.

b. The standard deviation is $\sqrt{\dfrac{1.5^2}{25} + \dfrac{2^2}{23}} = 0.514$.

c. The figure is centered at 3.5 and has a standard deviation of 0.514.

Figure for Exercise 9.79c

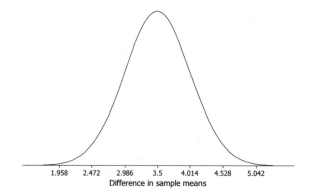

74

d. Notice in the figure in part c that a sample difference of 3.6 days falls just above the mean. It corresponds to a z-score of $(3.6 - 3.5)/0.514 = 0.19$, so it is reasonable.

9.81   **a.** The mean is $70 - 65 = 5$ inches.

     b. The standard deviation is $\sqrt{\dfrac{3^2}{9} + \dfrac{2.5^2}{9}} = 1.302$, round to 1.30.

     c. The picture is approximately normal with mean of 5 and standard deviation of 1.30.

Figure for Exercise 9.81c

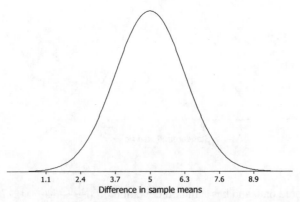

Difference in sample means

     d. The probability that the mean for women is greater than the mean for men is $P(\overline{x}_1 - \overline{x}_2 < 0)$, which is the area below 0 for the figure in part c. Notice in the figure in part c that 0 falls below the lowest value on the horizontal axis, which represents three standard deviations below the mean. In fact 0 corresponds to a z-score of $(0 - 5)/1.3 = -3.85$, and the area below that is about .00006. Therefore, it is possible but not likely for the mean for women to be greater than the mean for men.

9.83   The population mean μ and the population standard deviation σ must also be known. The relevant formula is $z = \dfrac{\overline{x} - \mu}{\sigma / \sqrt{n}}$.

9.85   **a.** The picture is centered on 0 and has standard deviation of $\sqrt{\dfrac{(.5)(.5)}{100} + \dfrac{(.5)(.5)}{100}} = .0707$. (The point two standard deviations above the mean, 0.141, is not shown in the figure below because there is no room to fit it next to the value of 0.13.)

Figure for Exercise 9.85a

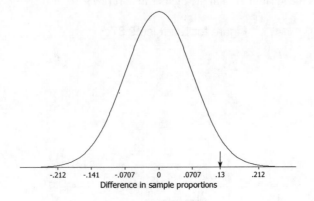

Difference in sample proportions

75

b. $(0.13 - 0)/.0707 = 1.84$

c. The new picture is centered on 0 and has standard deviation of 1. The location of 1.84 on this picture should be the same as the location of 0.13 in the picture in part (a).

Figure for Exercise 9.85c

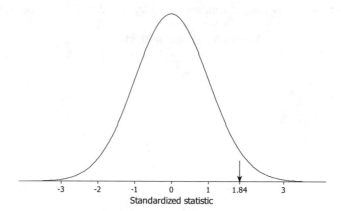

9.87   **a.** The picture is centered on 0 and has standard deviation of $\sqrt{\dfrac{3^2}{30} + \dfrac{3^2}{30}} = 0.775$.

Figure for Exercise 9.87a

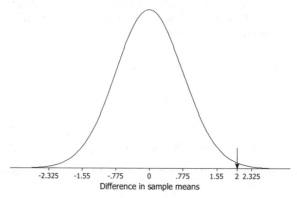

b. $(2 - 0)/0.775 = 2.58$.

c. The new picture is centered on 0 and has standard deviation of 1. The location of 2.58 on this picture should be the same as the location of 2 in the picture in part (a).

Figure for Exercise 9.87c

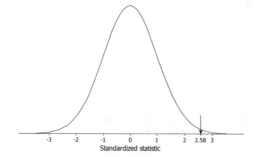

76

9.89   **a.** The sampling distribution is approximately normal with mean 0 and standard deviation $5/\sqrt{12} = 1.44$.
b. The standardized statistic is $(8 - 0)/1.44 = 5.56$. This is not a reasonable value. A difference resulting in a z-score of that magnitude would occur very rarely if in fact the population mean difference is 0.

9.91   **a.** Assuming the original populations are bell-shaped, the sampling distribution is approximately normal

with mean 0 and standard deviation of $\sqrt{\dfrac{1.5^2}{25} + \dfrac{2^2}{23}} = 0.5137$.

b. A sample difference of 3.6 days corresponds to a z-score of $(3.6 - 0)/0.5137 = 7.01$.
c. The observed result is not likely to be a statistical fluke. A standardized score of 7 or more would occur in fewer than 1.29×10^{-12} samples (found using Excel).

9.93   **a.** $z = \dfrac{\hat{p} - p}{s.d.(\hat{p})} = \dfrac{\hat{p} - p}{\sqrt{\dfrac{p(1-p)}{n}}} = \dfrac{.60 - .50}{\sqrt{\dfrac{.50(1-.50)}{100}}} = \dfrac{.10}{.05} = 2.0$.

b. $z = \dfrac{\hat{p} - p}{s.d.(\hat{p})} = \dfrac{\hat{p} - p}{\sqrt{\dfrac{p(1-p)}{n}}} = \dfrac{.60 - .50}{\sqrt{\dfrac{.50(1-.50)}{200}}} = \dfrac{.10}{.0354} = 2.828$.

9.95   **a.** For $\hat{p} = .20$, $z = \dfrac{\hat{p} - p}{s.d.(\hat{p})} = \dfrac{\hat{p} - p}{\sqrt{\dfrac{p(1-p)}{n}}} = \dfrac{.20 - .25}{\sqrt{\dfrac{.25(1-.25)}{75}}} = \dfrac{-.05}{.05} = -1$.

For $\hat{p} = .33$, $z = \dfrac{\hat{p} - p}{s.d.(\hat{p})} = \dfrac{.33 - .25}{\sqrt{\dfrac{.25(1-.25)}{75}}} = \dfrac{.08}{.05} = 1.6$.

b.

Figure for Exercise 9.95b

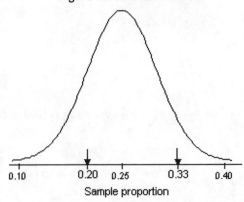

Sample proportion

c.

Figure for Exercise 9.95c

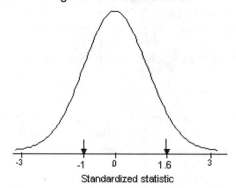

Standardized statistic

d. In essence, the drawings for parts (b) and (c) are the same. In part (c) the horizontal axis shows standardized statistics that correspond to the values along the horizontal axis in part (b). The area (probability) between $\hat{p} = .20$ and $\hat{p} = .33$ equals the area (probability) between $z = -1$ and $z = 1.6$.

9.97 **a.** $z = \dfrac{\bar{x} - \mu}{s.d.(\bar{x})} = \dfrac{\bar{x} - \mu}{\dfrac{\sigma}{\sqrt{n}}} = \dfrac{74 - 72}{\dfrac{10}{\sqrt{25}}} = \dfrac{2}{2} = 1$.

b. $z = \dfrac{\bar{x} - \mu}{s.d.(\bar{x})} = \dfrac{\bar{x} - \mu}{\dfrac{\sigma}{\sqrt{n}}} = \dfrac{70 - 72}{\dfrac{10}{\sqrt{25}}} = \dfrac{-2}{2} = -1$.

9.99 Mean $= \mu = \$200{,}000$; s.d. $(\bar{x}) = \dfrac{\sigma}{\sqrt{n}} = \dfrac{\$25{,}000}{\sqrt{25}} = \$5{,}000$; $z = \dfrac{208{,}000 - 200{,}000}{5000} = 1.6$.

9.101 **a.** $t = \dfrac{\bar{x} - \mu}{s.e.(\bar{x})} = \dfrac{\bar{x} - \mu}{\dfrac{s}{\sqrt{n}}} = \dfrac{5 - 10}{\dfrac{20}{\sqrt{16}}} = \dfrac{-5}{5} = -1$; df $= n - 1 = 16 - 1 = 15$.

b. $t = 1$; df $= 15$.

9.103 **a.** $t = \dfrac{\bar{x} - \mu}{s.e.(\bar{x})} = \dfrac{\bar{x} - \mu}{\dfrac{s}{\sqrt{n}}} = \dfrac{175 - 170}{\dfrac{24}{\sqrt{4}}} = \dfrac{5}{12} = 0.417$; it's t because s is used instead of σ.

b. $z = \dfrac{\bar{x} - \mu}{s.e.(\bar{x})} = \dfrac{\bar{x} - \mu}{\dfrac{\sigma}{\sqrt{n}}} = \dfrac{175 - 170}{\dfrac{20}{\sqrt{4}}} = \dfrac{5}{10} = 0.5$; it's z because σ is known.

c. $t = \dfrac{\bar{x} - \mu}{s.e.(\bar{x})} = \dfrac{\bar{x} - \mu}{\dfrac{s}{\sqrt{n}}} = \dfrac{161 - 170}{\dfrac{18}{\sqrt{36}}} = \dfrac{-9}{3} = -3$; it's t because s is used instead of σ.

9.105 The symbol s represents the standard deviation of a sample. The symbol σ represents the standard deviation of the population.

9.107 **a.** t-distribution with df $= 10$. Generally, a t-distribution is more spread out than a standard normal curve (although the difference is not great for larger values of degrees of freedom).

b. *t*-distribution with df = 5. Within the family of *t*-distributions, a *t*-distribution becomes less spread out as the degrees of freedom value increases.

c. Normal distribution with standard deviation = 100. A t-distribution is similar to a standard normal curve, which has standard deviation = 1.

9.109 **a.** $t = \dfrac{\bar{x}-\mu}{s.e.(\bar{x})} = \dfrac{\bar{x}-\mu}{\dfrac{s}{\sqrt{n}}} = \dfrac{78,000-80,000}{\dfrac{4,000}{\sqrt{100}}} = \dfrac{2,000}{400} = -5$. The symbol t is used because the sample standard deviation (rather than the population standard deviation) is used.

b. The degrees of freedom are df = n – 1 = 99, and the probability below *t* = –5 is approximately .000001 (1 in a million). If Excel is used to find the probability, it's necessary to the probability above t = +5 (which equals the probability below –5 due to symmetry). The command is TDIST(5,99,1).

Figure for Exercise 9.109b

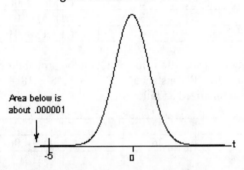

9.111 Yes, the value is 0. Every *t*-distribution is centered at 0, so the area above *t* = 0 is .5 no matter what the degrees of freedom.

9.113 **a.**

Sample	$\bar{x}$	Sample	$\bar{x}$
1,3	2	3,7	5
1,5	3	3,9	6
1,7	4	5,7	6
1,9	5	5,9	7
3,5	4	7,9	8

b.

$\bar{x}$	2	3	4	5	6	7	8
Probability	1/10	1/10	2/10	2/10	2/10	1/10	1/10

9.115 **a.**

Sample	R	Sample	R
1,2	2–1=1	2,4	4–2=2
1,3	3–1=2	2,5	5–2=3
1,4	4–1=3	3,4	4–3=1
1,5	5–1=4	3,5	5–3=2
2,3	3–2=1	4,5	5–4=1

b. 1/10 (only one way to have *R* =4).

c.

R	1	2	3	4
Probability	4/10	3/10	2/10	1/10

79

d. $R = 1$ is most likely; $R = 4$ is least likely.

9.117 **a.** For $T =$ total dogs in two households, possible values are 0, 1, 2, 3, and 4.
b. In the following table, every possible combination of the numbers of dogs in the two households is shown along with the corresponding value of T, and the calculation of the probability for that combination. The multiplication rule (from Chapter 7) is used to calculate each probability.

Dogs in first house	Dogs in second house	$T =$ total dogs in two houses	Probability
0	0	0	$(.6)(.6) = .36$
0	1	1	$(.6)(.3) = .18$
0	2	2	$(.6)(.1) = .06$
1	0	1	$(.3)(.6) = .18$
1	1	2	$(.3)(.3) = .09$
1	2	3	$(.3)(.1) = .03$
2	0	2	$(.1)(.6) = .06$
2	1	3	$(.1)(.3) = .03$
2	2	4	$(.1)(.1) = .01$

To find the overall probability for a specific value of T, add the probabilities associated for each combination having that value of T. For instance, $P(T = 1) = .18 + .18 = .36$. The distribution of T can be summarized as follows:

t	0	1	2	3	4
$P(T = t)$	.36	.36	.21	.06	.01

c. About $(1000)(.36) = 360$ teens will have $T = 0$ on their first try.
Note: In part (b), it was found that $P(T = 0) = .36$. With a relative frequency interpretation, this means that in the long run the fraction of times that $T = 0$ is about .36 (or 36%).

9.119 Because of the symmetry of the situation, the distribution of the lowest number should be a mirror image of the distribution given in Figure 9.13 for the highest number, so it should be highly skewed to the right. *Note*: The raw data are in the dataset **Cash5** on the companion website for the book, so this answer can be confirmed by drawing a histogram of the lowest numbers for the 1,560 plays.

9.121 **a.** Probability = 1/2 that any particular line will be "broken." To verify this, list the eight possible outcomes for the three pennies for each line, and notice that the number of heads is odd in one-half of the outcomes. The possible outcomes are HHH, TTH, HTT, THT, HHT, HTH, THH, and TTT.
b. B can be any of 0, 1, 2, 3, 4, 5, 6. Each of the six lines can be broken or not.

c. $P(B=0) = \left(\dfrac{1}{2}\right)^6 = \dfrac{1}{64}$. This is an application of the multiplication rule for independent events.

d. The sampling distribution of B is a binomial distribution with $n = 6$ and $p = 1/2$. The six lines are like six trials of a binomial experiment. The probabilities, which can be found using either the binomial formula (in Section 8.4) or software, are:
$$P(B = 0) = 1/64 = .016$$
$$P(B = 1) = 6/64 = .094$$
$$P(B = 2) = 15/64 = .234$$
$$P(B = 3) = 20/64 = .3125$$
$$P(B = 4) = 15/64 = .234$$
$$P(B = 5) = 6/64 = .094$$
$$P(B = 6) = 1/64 = .016$$

9.123 **a.** The population of possible net gains is highly skewed (not bell-shaped) and the sample size is small.
b. The mean of the sampling distribution of $\bar{x}$ is the population mean $\mu = -\$0.35$.

The standard deviation of $\bar{x}$ is $\dfrac{\sigma}{\sqrt{n}} = \dfrac{\$29.67}{\sqrt{10}} = \$9.38$.

9.125 **a.** X = number correct has a binomial distribution with $n = 10$ and $p = .5$.
The number of trials ($n = 10$ questions) is fixed in advance, there are two possible outcomes for each trial, trials are independent, and the probability of success ($p = .5$) is the same for each trial.
b. $\mu = np = (10)(.5) = 5$ correct questions.
c. $\sigma = \sqrt{np(1-p)} = \sqrt{10(.5)(1-.5)} = 1.58$. (See page 280 of the text for the formulas for this and part b.)

9.127 **a.** Approximately a normal curve.
b. Answers will vary, but should be roughly the same as the answer to part (c), i.e. 5.5 and 10.5.
c. Interval is $8 \pm (3 \times \dfrac{5}{\sqrt{36}})$, which is 8 ± 2.5, or 5.5 to 10.5.

9.129 **a.** Answers will vary.
b. Answers will vary.
c. There is less variation among sample means for the larger sample size ($n = 100$).

9.131 **a.** For sampling distribution, mean = 8.352 hours and standard deviation $= \dfrac{7.723}{\sqrt{49}} = 1.103$ hours.

Interval is $8.352 \pm (2 \times 1.103)$, or 6.146 to 10.558 hours. Roughly, about 6.1 to 10.6 hours.
b. Interval is 8.352 ± 1.103, or 7.249 to 9.455 hours. Roughly, about 7.2 to 9.5 hours.

9.133 The Rule for Sample Proportions applies because the three conditions are met. There is an actual population (the students at your college) and a fixed (although unknown) proportion of those students are left-handed. A random sample of 200 students at your college will be taken. The sample size, 200 is large enough so that both np and $n(1-p)$ will be greater than 10.

9.135 **a.** The distribution of possible sample proportions is approximately normal with mean=.50 and standard

deviation $s.d.(\hat{p}) = \sqrt{\dfrac{p(1-p)}{n}} = \sqrt{\dfrac{.5(1-.5)}{800}} = .018$

Figure for Exercise 9.135a

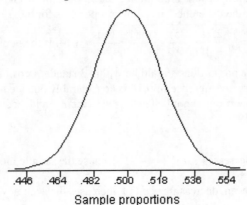

Sample proportions

b. Yes, a sample of 800 people from a population with a proportion of .50 would be unlikely to result in a sample proportion of .70, as can be seen in the Figure in part (a). Nearly all sample proportions will be within 3 standard deviation of .50, which is .50± (3)(.0177) or .447 to .553.

9.137   The sum of a list of 1s and 0s equals the number of 1s so the mean (sum divided by the sample size) will equal the proportion of 1s (number of 1s divided by sample size). This can be verified with any arbitrary list of 1s and 0s. For instance, for the list 0, 1, 0, 0, 1, the mean is 2/5 = 0.4, which is the proportion of 1s.

9.139   The mean still is 25 mpg, but the standard deviation is now $s.d.(\bar{x}) = \dfrac{\sigma}{\sqrt{n}} = \dfrac{1}{\sqrt{100}} = 0.1\,\text{mpg}$.

The normal curve will be much more tightly bunched around 25. The main idea is that a sample mean based on $n=100$ is likely to be closer to the true mean than a sample mean based on $n=9$.

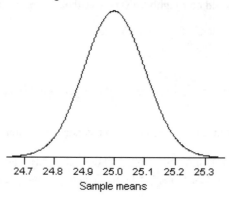

Figure for Exercise 9.139

9.141   The distribution will be approximately a normal curve with mean 105 and standard deviation $s.d.(\bar{x}) = \dfrac{\sigma}{\sqrt{n}} = \dfrac{15}{\sqrt{100}} = 1.5$.

9.143   **a.** Note that the standard error in 1976 is $s.e.(\hat{p}) = \sqrt{\dfrac{\hat{p}(1-\hat{p})}{n}} = \sqrt{\dfrac{.09(1-.09)}{8709}} = .003$ so the difference between the sample and population proportions is almost surely less than $3\,s.e.(\hat{p}) = (3)(.003) = .009$ (or 0.9% or about 1%).
b. The difference in sample proportions (subtracting the 1976 result from the 1997 result) is $.25 - .09 = 0.16$. The standard error for the difference in sample proportions for these sample sizes is

$$\sqrt{\dfrac{\hat{p}_1(1-\hat{p}_1)}{n_1} + \dfrac{\hat{p}_2(1-\hat{p}_2)}{n_2}} = \sqrt{\dfrac{.25(1-.25)}{1000} + \dfrac{.09(1-.09)}{8709}} = .014.$$ If the population proportions were equal

then the difference in sample proportions should be within 3 standard errors of 0, which is the interval from -0.042 to $+0.042$. The observed difference of 0.16 is considerably outside of this range, indicating that the difference in population proportions is not 0. There is convincing evidence that the population proportion was higher in 1997 than in 1976.

9.145   Increasing the sample size decreases $s.d.(\bar{x}) = \dfrac{\sigma}{\sqrt{n}}$. Because the sample mean is likely to be within 3

$s.d.(\bar{x})$ of the population mean, decreasing $s.d.(\bar{x})$ increases the accuracy of the sample mean as an estimate of the population mean.

9.147   **a.** No, the conditions are not met. The sample size is not large enough. If $p = .10$ and $n = 30$, then $np = 3$, which is less than 10.
b. Yes, assuming that the proportion of interest is a long-run relative frequency over all weekdays and seasons, and the days on which they do the survey is representative of all days and seasons. The fixed probability is that someone will be home during those hours at a randomly selected residence on a randomly selected day. The sample size is large enough.

82

c. No, the conditions are not met. A random sample of days of the year was not taken, since the weather was recorded only for days in January and February. Clearly, snow or rain (depending on the area) will occur more frequently in those months than for the year as a whole.

d. Yes the conditions are met. The population consists of all employees of the company and a fixed proportion p of those employees is currently interested in on-site day care. A random sample was taken and the sample size of 100 is large enough unless p is very close to 0 or 1, which is not likely.

9.149 **a.** With a sample of $n=1600$ and a true proportion of $p= .56$, the standard deviation is

$$s.d.(\hat{p}) = \sqrt{\frac{56(1-.56)}{1600}} = .0124$$

About 68% of all potential sample proportions are in the range .56 ±.0124 or .5476 to .5724.
About 95% of all potential sample proportions are in the range .56 ± (2 ×.0124) or .535 to .585.
Almost always, the sample proportion will be in the range .56 ± (3 ×.0124) or .523 to .597.

b. Based on part (a), .61 does not seem like a reasonable sample proportion because it is outside the range given for the sample proportions that should occur almost always. The standardized score for .61, if the true proportion is .56, is (.61 - .56)/.0124 = 4.03.

c. For $n=400$, $\sqrt{\frac{56(1-.56)}{400}} = .025$. The reported percentage of 61% then corresponds to a standardized score of (.61 - .56)/.025 = 2. Only 2.5% of the potential sample proportions would have standardized scores larger than 2 if the true proportion is .56. If everybody told the truth, the sample result is unusual, but not nearly as unusual as with the sample size of 1600. This example illustrates the role of the sample size is assessing whether a sample result is inconsistent with a potential population value.

9.151 The phrase "almost sure": is vague but has been used to refer to the range mean ± 3*s.d* and will be interpreted that way. The administration wants $3 \times s.d.(\overline{x}) = 1$. This means that $3\dfrac{\sigma}{\sqrt{n}} = 3\dfrac{5}{\sqrt{n}} = 1$ which leads to $\sqrt{n} = 15$ and $n = 225$.

9.153 **a.** The method used will depend upon available software.
b. Here is the histogram resulting from one such simulation.

Figure for Exercise 9.153b

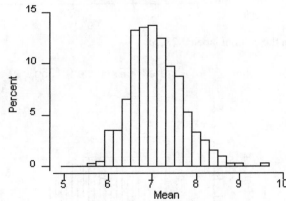

c. The distribution is skewed somewhat to the right. This happens because there are two outliers (16 hours of sleep) in the "population" of 190 students and when one or both of these observations are in a sample of $n=10$, the sample mean will be relatively large. If the outliers were removed from the population, the distribution of the sample mean would be approximately bell-shaped.
d. Here is the histogram resulting from the same simulation done for part (b).

Figure for Exercise 9.153d

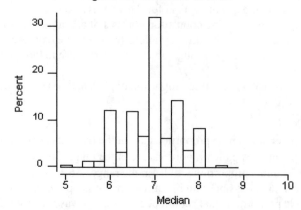

e. The outliers do not affect the median, so the sample median is almost never higher than 8 hours. Many observations in the "population" equal 7 hours and that probably leads to the high percentage of times that the sample median is exactly 7 hours.

9.155   **a.** The histogram is similar to Figure 9.13, as it should be. Both figures (this one and Figure 9.13) are estimates of the same theoretical sampling distribution.

Figure for Exercise 9.155a

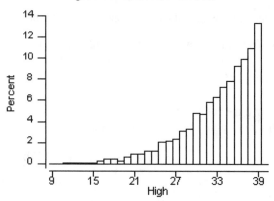

b. The median is in the column labeled *Third*.

c.

Figure for Exercise 9.155c

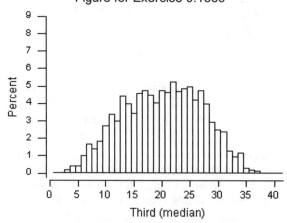

84

d. If using Minitab, use **Calc>Row Statistics** to calculate the means. Click "Mean" , enter the five columns containing the data as "Input Variables" and give an unused column for "Store result in:"

Figure for Exercise 9.155d

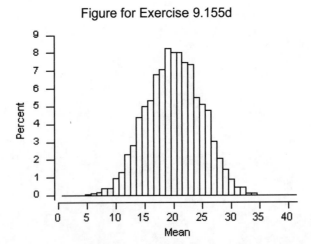

e. Both histograms are centered at 20 (which is the average of 1 through 39), but the histogram for the median is more spread out. The distribution of the sample mean is less spread out because the mean utilizes numerical information from all 5 observations (rather than just one), and so it is a more precise estimate of the center of the distribution than the median is.

CHAPTER 10
ODD-NUMBERED SOLUTIONS

10.1   **a.** This is a population proportion.
b. This is a sample proportion.
c. This is a sample proportion. Actually, a sample percent is given in the problem. The corresponding proportion is .55.

10.3   **a.** How do American teenagers rate their parents on strictness, compared to their friends' parents? In particular, what proportion thinks that their parents are more strict than their friends' parents?
b. The parameter of interest is the proportion (or percentage) of American teenagers who think their parents are more strict than their friends' parents.
c. 39%, found as 100% × 171/439.
d.

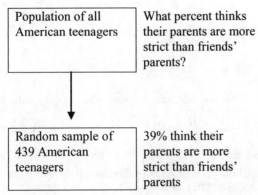

10.5   100% − 95% = 5%. With a 95% confidence level, about 5% of all random samples from a population will provide a confidence interval that does not cover the population value.

10.7   **a.** All U.S. adults
b. Proportion of U.S. adults who think that the use of handheld cell phones while driving should be illegal
c. The $n = 836$ who were surveyed
d. .72 (or 72%)

10.9   **a.** The margin of error is given as 3.5%. Therefore, a 95% confidence interval is 36% ± 3.5% which is 32.5% to 39.5%. With 95% confidence, we can say that in the population of American adults, between 32.5% and 39.5% would say they don't get enough sleep each night.
b. No. Using the Fundamental Rule for Using Data for Inference, the sample could be used in this way only if American adults are representative of college students for this question. If college students generally differ from others on this question, which is highly likely, then this sample won't correctly estimate the percent for students.

10.11   This sample should not be used to calculate a confidence interval because it is a self-selected volunteer sample and such samples are nearly always biased toward a certain viewpoint. In this case, it is possible that people who have been victims or know victims may be more likely to respond so the sample percentage will overestimate the population percentage. Therefore, the Fundamental Rule for Using Data for Inference doesn't hold.

10.13   About 190, which is 95% of the 200 computed intervals, should contain the population proportion and about 10 should not. Remember that the confidence level describes the confidence in the procedure and gives the percentage of all intervals that will contain the true value.

10.15 **a.** Increasing the confidence level will increase the width of the interval.
      b. Decreasing the confidence level will decrease the width of the interval.

10.17 **a.** p = the proportion of the population that suffers from allergies.
      b. p = the proportion of the population of working adults who say that they would quit their jobs if they won a large amount in the lottery.
      c. Define μ_1 to be the mean severity score for premenstrual symptoms in the population if all women with premenstrual symptoms were to take calcium, and μ_2 to be the mean severity score if everyone in the population were to take a placebo. The parameter of interest is $\mu_1 - \mu_2$.

10.19 **a.** $s.e.(\hat{p}) = \sqrt{\dfrac{\hat{p}(1-\hat{p})}{n}} = \sqrt{\dfrac{.59(1-.59)}{1000}} = .016$

      b. The 95% confidence interval is $\hat{p} \pm 2\, s.e.(\hat{p})$ which is $.59 \pm (2 \times .016)$ or .558 to .622.
      With 95% confidence, we can say that between .558 and .622 of American adults think the world will come to an end.

10.21 (ii) $n = 400$, confidence level = 95%; the smaller sample size and greater confidence level both will tend to increase the width of the confidence interval

10.23 **a.** The sample proportion is $\hat{p} = 166/200 = .83$.
      b. The "formula" to use is *Sample estimate* $\pm$ *Multiplier* $\times$ *Standard error*. The sample estimate is $\hat{p} = .83$ and the standard error is $\sqrt{\dfrac{\hat{p}(1-\hat{p})}{n}} = \sqrt{\dfrac{.83(1-.83)}{200}} = .0266$. For 95% confidence the multiplier is 1.96. The 95% confidence interval is $.83 \pm (1.96 \times .0266)$, which is .778 to .882.
      c. With 95% confidence, we can say that if the whole population with this disease received this treatment, the proportion successfully treated would be between .778 and .882 (or 77.8% to 88.2%).

10.25 The "formula" to use is *Sample estimate* $\pm$ *Multiplier* $\times$ *Standard error*. The sample estimate is $\hat{p} = .83$ and the standard error is $s.e.(\hat{p}) = \sqrt{\dfrac{\hat{p}(1-\hat{p})}{n}} = \sqrt{\dfrac{.83(1-.83)}{200}} = .0266$. For 98% confidence the multiplier given in Table 10.1 is $z^* = 2.33$. The 98% confidence interval is $.83 \pm (2.33 \times .0266)$, which is $.83 \pm .062$ or .768 to .892. This interval is wider than the interval for the 95% confidence level given in part b of Exercise 10.19.

10.27 **a.** $\hat{p} = \dfrac{220}{400} = .55$.

      b. $s.e.(\hat{p}) = \sqrt{\dfrac{\hat{p}(1-\hat{p})}{n}} = \sqrt{\dfrac{.55(1-.55)}{400}} = .025$

      c. $.55 \pm (1.96)(.025)$, which is $.55 \pm .049$, or .501 to .599. (You could round the multiplier to 2, resulting in an interval from .50 to .60.)
      d. $.55 \pm (2.33)(.025)$, which is $.55 \pm .058$, or .492 to .608. $z^* = 2.33$ is the multiplier for 98% confidence (use Table 10.1 to find this).

10.29 **a.** All adults in the United States
      b. $\sqrt{\dfrac{\hat{p}(1-\hat{p})}{n}} = \sqrt{\dfrac{.60(1-.60)}{537}} = .0211$
      c. $.60 \pm 1.96\,(.0211)$, which is about .559 to .641
      d. With 95% confidence we estimate that between .559 and .641 of all adults in the United States think that penalties for underage drinking should be stricter.

10.31 A necessary condition for using the methods of this chapter is that the number in each category is at least 10 (although some authors say at least 5), and there is only one left-handed person in the sample. *Note*: Theoretically, the sample size condition has to do with the expected numbers in the categories. Only slightly more than 10% of humans are left-handed, so any sample this small will almost certainly have fewer than 10 left-handed people. The expected number of left-handed people in a sample of $n = 15$ is only about 1.5, many fewer than the 10 (or 5) required to use the procedures in this Chapter.

10.33 The "formula" to use is *Sample estimate $\pm$ Multiplier $\times$ Standard error*. The sample estimate is $\hat{p} = 57/300 = .19$ and the standard error is $s.e.(\hat{p}) = \sqrt{\dfrac{\hat{p}(1-\hat{p})}{n}} = \sqrt{\dfrac{.19(1-.19)}{300}} = .0226$. For 90% confidence the multiplier given in Table 10.1 is $z^* = 1.645$. The 90% confidence interval is $.19 \pm (1.645 \times .0226)$, which is $.19 \pm .037$ or .153 to .227. We are 90% confident that in the population of employed Americans the proportion who would fire their boss if they could is between .153 and .227.

10.35 $z^* = 1.28$. To find this, use the standard normal curve to find the value z^* such that the probability between $-z^*$ and $+z^*$ is .80. If .80 is the probability between $-z^*$ and $+z^*$, then .10 is the probability to the left of $-z^*$. And, .90 is the probability to the left of $+z^*$ (because .10 is to the right of this value). In Table A.1, look either for .10 or for .90 where the probabilities are given, and determine the corresponding z-value.

10.37 .80, or 80%

10.39 The multiplier is $0..67$ (see Example 10.5 on page 382 for details). The 50% confidence interval is $.60 \pm 0.67\sqrt{\dfrac{.60(1-.60))}{537}}$, which is $.60 \pm (0.67)(.0211)$, or about .586 to .614.

10.41 **a.** 99% of the time.
b. 1% of the time.

10.43 About $100\% - 95\% = 5\%$ of the time, the difference between the sample percentage and the population percentage will be greater than 3%. This is assuming that, as usual, the margin of error is associated with 95% confidence.

10.45 **a.** $2\sqrt{\dfrac{\hat{p}(1-\hat{p})}{n}} = 2\sqrt{\dfrac{.56(1-.56)}{100}} = .0993$.

b. $2\sqrt{\dfrac{\hat{p}(1-\hat{p})}{n}} = 2\sqrt{\dfrac{.56(1-.56)}{400}} = .0496$.

10.47 **a.** $s.e.(\hat{p}) = \sqrt{\dfrac{\hat{p}(1-\hat{p})}{n}} = \sqrt{\dfrac{.03(1-.03)}{883}} = .00574$

b. $2\,s.e.(\hat{p}) = 2(.00574) = .01148$, or 1.1%.

10.49 **a.** $\dfrac{1}{\sqrt{n}} = \dfrac{1}{\sqrt{757}} = .036$, which verifies that the margin of error is about 3.5%.

b. A 95% confidence interval is $3\% \pm 3.5\%$ or -0.5% to 6.5%. Not all of these values are valid estimates of the population percentage since the population percentage cannot be less than 0.

c. The standard error is $s.e.(\hat{p}) = \sqrt{\dfrac{\hat{p}(1-\hat{p})}{n}} = \sqrt{\dfrac{.03(1-.03)}{757}} = .006$.

The margin of error is $2 \times standard\ error = 2\,(.006) = .012$ or 1.2%.

d. A 95% confidence interval based on the margin of error in part c is 3% ± 1.2% or 1.8% to 4.2%. This means that it is highly likely that the percentage of all American women who would say they are "Not at all satisfied" with their overall physical appearance is between 1.8% and 4.2%.

10.51 **a.** Yes, there are two independent samples with sufficiently large sample sizes.
b. No, the sample sizes are too small.
c. No, The methods of Section 10.4 should not be used because the analysis involves means, not proportions.

10.53 This is the difference between the proportions in two categories of a variable within the same single sample. The method in Section 10.4 is for the difference in the proportions with the same trait in two independent samples.

10.55 **a.** Age 30-39: $\hat{p}_1 = \dfrac{995}{2122} = .469$

Age 18-29: $\hat{p}_2 = \dfrac{653}{1600} = .408$

Difference is $\hat{p}_1 - \hat{p}_2 = .469 - .408 = .061$

b. Approximate 95% interval is .028 to .094, computed as $.061 \pm (2)(.0164)$. With $z^* = 1.96$ as the multiplier, the answer is .029 to .093.
Parameter is $p_1 - p_2$ = difference in proportions experiencing episodic tension-type headaches in the populations of women aged 18 to 29 and women aged 30 to 39
Compute the interval as Sample estimate ± Multiplier × Standard error:
Sample estimate is $\hat{p}_1 - \hat{p}_2 = .061$

Standard error is: $s.e.(\hat{p}_1 - \hat{p}_2) = \sqrt{\dfrac{\hat{p}_1(1-\hat{p}_1)}{n_1} + \dfrac{\hat{p}_2(1-\hat{p}_2)}{n_2}} = \sqrt{\dfrac{.469(1-.469)}{2122} + \dfrac{.401(1-.401)}{1600}} = .0164$

Multiplier = $z^* \approx 2$ (or 1.96 is more exact)
Interpretation: With 95% confidence, we can say that the difference in proportions experiencing episodic tension-type headaches in the populations of women aged 18 to 29 and women aged 30 to 39 is between .028 and .094. The proportion is higher for women aged 30 to 39.
Note: Minitab could be used to find the interval. (See Tip on p.426.)

10.57 **a.** *Interpretation*: With 95% confidence, we can say that the difference in proportions experiencing episodic tension-type headaches in the populations of women with at least a college degree and at least a high school degree is between 0.068 and 0.104. The proportion is higher for women with at least a college degree. The interval dos not cover 0 so it is reasonable to conclude that the proportions differ in the populations.
b. Sample estimate ± Multiplier × Standard error
Sample estimate = $\hat{p}_1 - \hat{p}_2 = .466 - .380 = .086$

Standard error $= s.e.(\hat{p}_1 - \hat{p}_2) = \sqrt{\dfrac{\hat{p}_1(1-\hat{p}_1)}{n_1} + \dfrac{\hat{p}_2(1-\hat{p}_2)}{n_2}} = \sqrt{\dfrac{.466(1-.466)}{4594} + \dfrac{.380(1-.380)}{7076}}$

Multiplier = $z^* = 1.96$ (which might be rounded to 2)

10.59 **a.** $\hat{p}_1 - \hat{p}_2 = .70 - .40 = .30$.

b. $s.e.(\hat{p}_1 - \hat{p}_2) = \sqrt{\dfrac{.70(1-.70)}{100} + \dfrac{.40(1-.40)}{80}} = .0714$.

c. Approximate 95% confidence interval is $.30 \pm (2 \times 0.0714)$, or about .157 to .443.
d. Yes. The confidence interval does not cover 0.

10.61 **a.** 55% ± 3%, or 52% to 58%. This is calculated as *Sample estimate ± Margin of error*.

90

b. Yes. All of the values in the 95% confidence interval are greater than 50%, so it's reasonable to infer that more than 50% of the population thinks their weight is about right.

10.63   The claim of 28% is not reasonable because the value 28% not in the confidence interval.

10.65   **a.** 95% confidence intervals for the two years do not overlap. It's reasonable to conclude that the population percent is higher in 2002. The 95% confidence intervals can be computed as *Sample percent ± Margin of error*. For 1999, the confidence interval is 46% ± 3%, or 43% to 49%. For 2002, the confidence interval is 55% ± 3%, or 52% to 58%.
b. Define population 1 to be 2002 and population 2 to be 1990. Then $\hat{p}_1 - \hat{p}_2 = .55 - .46 = .09$,

$$s.e.(\hat{p}_1 - \hat{p}_2) = \sqrt{\frac{.55(.45)}{1004} + \frac{.46(.54)}{1004}} = 0.022,$$ and the 95% confidence interval is $.09 \pm 2(.022)$, which is 0.046 to 0.134.

c. We can be 95% confident that the proportion of U.S. adults who felt that their weight is about right increased from 1990 to 2002 and that the increase in the number who felt that way was somewhere between 0.046 (4.6%) and 0.134 (13.4%) of the population.

10.67   **a.** The population is all University of California faculty members in 1995.

b. The margin of error is approximately $\frac{1}{\sqrt{n}} = \frac{1}{\sqrt{1000}} = .032$, or roughly 3%.

c. It cannot be concluded that a majority of all University faculty favored the criteria. In the sample a slight majority (52%) was in favor, but the margin of error was about 3%. So, in the population the percent in favor could possibly be a minority (below 50%). An interval that is 95% certain to contain the percent in the population is 52% ± 3%, which is 49% to 55%. Because some values within this interval are below 50%, we are not able to rule out the possibility that the percent in favor in the population is a minority.

10.69   **a.** The sample proportion should not be used to estimate the population proportion because the sample was a self-selected sample. Such samples usually do not represent any larger population.
b. While the sample is not a random sample, for the question of interest it may provide a suitable estimate of the proportion that is left-handed in the population. It doesn't seem that this sample of 400 students should be biased toward having a different proportion that's left-handed than in the population.

10.71   **a.** $\hat{p} = \frac{56}{190} = .295$

b. $s.e.(\hat{p}) = \sqrt{\frac{.295(1-.295)}{190}} = .033$

c. For 90% confidence, the multiplier is $z^* = 1.645$ (see Table 10.1). The interval is $\hat{p} \pm z^* s.e.(\hat{p})$ which is $.295 \pm (1.645 \times .033)$. This is $.295 \pm .054$, or .241 to .349.
d. For 95% confidence, the multiplier is approximately $z^* = 2$ (see Table 10.1). The interval is $\hat{p} \pm z^* s.e.(\hat{p})$ which is $.295 \pm (2 \times .033)$. This is $.295 \pm .066$, or .229 to .361.

e. For 98% confidence, the multiplier is $z^* = 2.33$ (see Table 10.1). The interval is $\hat{p} \pm z^* s.e.(\hat{p})$ which is $.295 \pm (2.33 \times .033)$. This is 0.295 ± 0.077, or .218 to .372.
f. As the confidence level is increased, the width of the interval increases.
g. If numbers are chosen randomly, the true proportion who pick the number "7" would be 1/10=.1. None of the intervals calculated in the parts (c-e) include .1, so with any of those confidence levels, it's reasonable to conclude that the student picks are not made randomly.

10.73   The sample estimate is $\hat{p}_1 - \hat{p}_2 = .32 - .18 = 0.14$. The standard error is $s.e.(\hat{p}_1 - \hat{p}_2) =$

$\sqrt{\frac{.32(.68)}{232} + \frac{.18(.82)}{817}} = 0.03345.$ A 95% confidence interval for the difference in heart attack proportions

91

for the populations (policemen – other) is $0.14 \pm 1.96(.03345)$ or 0.15 ± 0.066 or 0.084 to 0.216. Yes, it is safe to conclude that police officers are more likely to experience heart disease than other men because the interval does not cover 0.

10.75 The professor's class is the entire population of interest and is not a sample from a larger group. A confidence interval is unnecessary because the population value is observed and does not have to be estimated.

10.77 **a.**

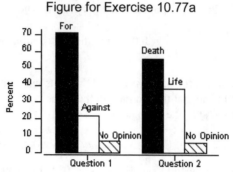

Figure for Exercise 10.77a

b. $\dfrac{1}{\sqrt{n}} = \dfrac{1}{\sqrt{543}} = 0.043$ while $\dfrac{1}{\sqrt{511}} = 0.044$. The Gallup Organization often rounds up, and this is probably why they reported the margin of error to be about 5%. *Note*: When rounding to two digits it is appropriate to round up because it is better to have an interval that is slightly wider than required for the stated confidence level than one that is slightly too narrow.

c. 71% ± 5% or 66% to 76%. We can be 95% confident that somewhere between 66% and 76% of all adult Americans would answer that they are for the death penalty when asked Question 1.

d. 56%± 5% which is 51% to 61%. We can be 95% confident that between 51% and 61% of all adult Americans would respond "death penalty" when asked Question 2. Notice that all values in this interval are lower than all values in the interval of part (c). This is an example of how the wording of a question can affect how people respond.

10.79 **a.** The proportion of women with a tattoo = 46/273 = .168
b. The proportion of men with a tattoo = 32/207 = .155
c.

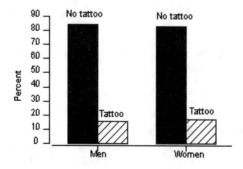

Figure for Exercise 10.79c

d. Ideally the results could be generalized to all college students in the U.S., but it might be safer to claim they can be generalized to all college students with liberal arts majors in the northeast. Also, tattoos go in and out of fashion, so the population could be further refined to include students in that geographic area in the late 1990's only.

92

e. The standard error is $s.e.(\hat{p}) = \sqrt{\dfrac{\hat{p}(1-\hat{p})}{n}} = \sqrt{\dfrac{.168(1-.168)}{273}} \approx .023$. A 95% confidence interval for the

college women is $.168 \pm (2 \times .023)$, which is .122 to .214. We have 95% confidence that in the population of college women represented by this sample, the proportion with tattoos is between .122 and .214.

f. The standard error is $s.e.(\hat{p}) = \sqrt{\dfrac{\hat{p}(1-\hat{p})}{n}} = \sqrt{\dfrac{.155(1-.155)}{207}} = .025$. A 95% confidence interval for the

college men is $.155 \pm (2 \times .025)$ or .105 to .205. We have 95% confidence that in the population of college men represented by this sample, the proportion with tattoos is between .105 and .205.

g. The intervals overlap substantially so we should not conclude there is a difference between the population proportions with a tattoo.

10.81 **a.** Approximate 95% confidence interval is .264 to .416, computed as $.34 \pm 2 \times .038$.

Parameter is $p_1 - p_2 =$ difference in proportions of men (population 1) and women (population 2) who have driven after having too much to drink

The interval is Sample estimate $\pm$ 2 $\times$ Standard error, which is $\hat{p}_1 - \hat{p}_2 \pm 2 \times \text{s.e.}(\hat{p}_1 - \hat{p}_2)$

Sample estimate $= \hat{p}_1 - \hat{p}_2 = .63 - .29 = .34$

Standard error $= \text{s.e.}(\hat{p}_1 - \hat{p}_2) = \sqrt{\dfrac{\hat{p}_1(1-\hat{p}_1)}{n_1} + \dfrac{\hat{p}_2(1-\hat{p}_2)}{n_2}} = \sqrt{\dfrac{.63(1-.63)}{300} + \dfrac{.29(1-.29)}{300}} = .038$

b. With 95% confidence, we can say the interval .264 to .416 covers the difference between the proportions of men and women in the population who would say they have driven after having too much to drink. The sample proportion for men was higher, so we estimate (with 95% confidence) the proportion for men in the population is somewhere between .264 and .416 above the proportion for women.

10.83 Approximate 95% confidence interval is .574 to .686, computed as $.63 \pm 2 \times .028$.

Parameter is $p =$ proportion who have driven after having too much to drink in the population of men

Interpretation: With 95% confidence, we can say that in the population of men represented by the sample, the proportion who would say they have driven after having too much to drink is between .574 and .686. The interval is Sample estimate $\pm$ 2 $\times$ Standard error.

Sample estimate is $\hat{p} = .63$ and standard error is $s.e.(\hat{p}) = \sqrt{\dfrac{\hat{p}(1-\hat{p})}{n}} = \sqrt{\dfrac{.63(1-.63)}{300}} = .028$

10.85 **a.** A 90% confidence interval is about .077 to .126, computed as $.1014 \pm (1.645)(.0147)$.

Parameter is $p =$ proportion of men with a tattoo in the population of Penn State men with no ear pierce.
The formula is Sample estimate $\pm$ Multiplier $\times$ Standard error:

Sample estimate is $\hat{p} = \dfrac{43}{424} = .1014$.

Standard error is $s.e.(\hat{p}) = \sqrt{\dfrac{\hat{p}(1-\hat{p})}{n}} = \sqrt{\dfrac{(.1014)(1-.1014)}{424}} = .0147$

Multiplier is $z^* = 1.645$. Use Table 10.1 on page 378.
Check necessary conditions: The sample size in each category is large enough so that this method can be used. It was assumed the sample represents a random sample of Penn State men.

b. A 90% confidence interval is about .235 to .361, computed as $.2979 \pm (1.645)(.0385)$ Parameter is $p =$
proportion of men with a tattoo in the population of Penn State men with no ear pierce.
For computing Sample estimate $\pm$ Multiplier $\times$ Standard error:

Sample proportion is $\hat{p} = \dfrac{42}{141} = .2979$.   Standard error is $\sqrt{\dfrac{(.298)(1-.298)}{141}} = .0385$

Multiplier is same as for part (a).
Check necessary conditions: The sample size in each category is large enough so that this method can be used. It was assumed the sample represents a random sample of Penn State men.

93

c. A 99% confidence interval is about 0.090 to 0.302, computed as $0.1965 \pm (2.576)(0.0412)$.

Parameter is $p_1 - p_2 =$ the difference in proportions of men with a tattoo for populations of Penn State men with an ear pierce (group 1) and without an ear pierce (group 2).

For computing Sample estimate $\pm$ Multiplier $\times$ Standard error:

Sample estimate is $\hat{p}_1 - \hat{p}_2 = .2979 - .1014 = 0.1965$

Standard error is $\text{s.e.}(\hat{p}_1 - \hat{p}_2) = \sqrt{\dfrac{.298(1-.298)}{141} + \dfrac{.101(1-.101)}{424}} = 0.0412$

Multiplier $= z^* = 2.576$. Use Table 10.1 on page 378.

Check necessary conditions: The necessary checks were done in parts (a) and (b).

Interpretation: We are 99% confident that in the population(s) of Penn State men represented by the sample(s), the difference in the proportions with a tattoo for men with an ear pierce versus men without an ear pierce is between 0.090 and 0.302. In other words, the percentage with a tattoo among men with ear pierce is between 9% and 30% above the percentage with a tattoo among men with no ear pierce.

10.87   **a.** Fish oil: $\hat{p}_1 = \dfrac{9}{14} = .6429$; Placebo: $\hat{p}_2 = \dfrac{3}{16} = .1875$;

Difference $= \hat{p}_1 - \hat{p}_2 = .6429 - .1875 = .4554$

$\text{s.e.}(\hat{p}_1 - \hat{p}_2) = \sqrt{\dfrac{.6429(1-.6429)}{14} + \dfrac{.1875(1-.1875)}{16}} = .161$

b. The necessary sample size conditions are not met here because $n_1\hat{p}_1 = 9$, $n_1(1-\hat{p}_1) = 5$ and $n_2\hat{p}_2 = 3$ are all less than 10. These are the numbers in the fish oil group who had and had not responded favorably, and the number in the placebo group who responded favorably.

c. If the experiment is "blind," the participants would not know the specific treatments (fish oil or placebo) they were assigned. If the experiment is "double-blind," neither the participants nor the researcher evaluating the outcome would know the specific treatment assignments. The term "blind" would also apply if the participants knew what they were taking (for instance if the fish oil had a distinctive taste) but the researcher evaluating the outcome did know who took which treatment.

10.89   **a.** The first step is to tally the number and percent of Democrats. Output 1 for this exercise is a tally done with Minitab (using **Stat>Tables>Tally**). The result is that 548 (34.31%) of 1,597 respondents said they preferred the Democratic Party.

Note: At the bottom of the output, $*= 13$ means that 13 people did not answer this question.

Output 1 for Exercise 10.89		
polparty	Count	Percent
Democrat	721	35.87
Independent	746	37.11
Other	38	1.89
Republican	505	25.12
N=	2010	
*=	13	

Next, calculate the confidence interval. It may be possible to use software (Minitab version 12 or higher will do the work), or you might have to do calculations "by hand." Output 2 is Minitab output (see page 380 of the text for guidance).

Output 2 for Exercise 10.89				
Sample	X	N	Sample p	95% CI
1	721	2010	0.358706	(0.337739, 0.379674)

Using the normal approximation.

To 3 decimal places, the 95% confidence interval is .338 to .380.

The "by hand" calculations, done to 3 decimal places are:

$\hat{p} = .359$

$$s.e.(\hat{p}) = \sqrt{\frac{\hat{p}(1-\hat{p})}{n}} = \sqrt{\frac{.359(1-.359)}{2010}} = .0107$$

$\hat{p} \pm 1.96 \times s.e.(\hat{p})$ is $.359 \pm (1.96 \times .0107)$, which is $.359 \pm .021$, or $.338$ to $.380$.

b. We are 95% confident that the proportion of U.S. adults who say they are Democrats (in 2008) is between .359 and .380.

10.91   **a.** $\hat{p} = 104/168 = .619$.

b. The "by hand" calculation is $.619 \pm 1.96\sqrt{\dfrac{.619(1-.619)}{168}}$ or .546 to .693. Minitab output is shown below (with blank lines removed).

Minitab Output for Exercise 10.91b
Test and CI for One Proportion: atfirst
``
Test of p = 0.5 vs p not = 0.5
Event = yes
Variable   X   N   Sample p   95% CI   Z-Value P-Value
atfirst   104 168 0.619048 (0.545615, 0.692481)   3.09   0.002
``

c. With 95% confidence, we can say that in the population represented by this sample, the proportion believing in love at first sight is between .546 and .692.

10.93   **a.** First, create a two-way table for the two variables to determine relevant counts. Of $n_1 = 462$ Democrats, the number owning a gun is 119 so $\hat{p}_1 = \dfrac{119}{462} = .2576$. Of $n_2 = 341$ Republicans, the number owning a gun is 169 so $\hat{p}_2 = \dfrac{169}{341} = .4956$ The 95% confidence interval for $p_1 - p_2$ is about -0.304 to -0.172. Minitab will compute the interval directly without finding the two-way table first. You must first create a subset of the worksheet containing only Democrats and Republicans. Results are shown in the output below.

Minitab Output for Exercise 10.93
Test and CI for Two Proportions: owngun, polparty
``
Event = Yes
polparty   X   N   Sample p
Democrat   119 462 0.257576
Republican   169 341 0.495601
Difference = p (Democrat) - p (Republican)
Estimate for difference: -0.238025
95% CI for difference: (-0.304404, -0.171647)
``

b. The confidence interval does not cover 0 so is evidence of difference in the population).

CHAPTER 11
ODD-NUMBERED SOLUTIONS

11.1   Are you asking a question about children in this school only, or using them to represent a larger population?

11.3   **a.** Procedure for two independent samples.
      b. Procedure for paired data.
      c. Procedure for two independent samples.

11.5   **a.** Standard error is $s.e.(\bar{x}) = \dfrac{s}{\sqrt{n}} = \dfrac{15}{\sqrt{100}} = 1.5.$

      b. Standard error is $s.e.(\bar{d}) = \dfrac{s_d}{\sqrt{n}} = \dfrac{21}{\sqrt{49}} = 3.$

      c. Standard error is $s.e.(\bar{x}_1 - \bar{x}_2) = \sqrt{\dfrac{s_1^2}{n_1} + \dfrac{s_2^2}{n_2}} = \sqrt{\dfrac{100}{50} + \dfrac{100}{50}} = 2.$

11.7   Table A.2 or appropriate software can be used.
      a. 2.13
      b. 1.70
      c. df = 9; t* = 2.26

11.9   **a.** The parameters of interest are the population means μ for Country A, μ for Country B, and the difference in population means for the two countries, denoted as $\mu_1 - \mu_2$.
      b. $\bar{x}_1 - \bar{x}_2 = 22.5 - 16.3 = 6.2$ days.

11.11   **a.** Are the mean levels of B cells the same for the populations of autistic and non-autistic children? If not, by how much do the mean levels differ?
      b. The parameter of interest is $\mu_1 - \mu_2$. where μ_1 is the mean level of B cells for the population of autistic children and μ_2. is the mean level of B cells for the population of non-autistic children. (You could define them in the reverse order.)
      c. Does $\mu_1 - \mu_2. = 0$? If not, what is its value?

11.13   **a.** Paired data would have to be used. The height measurements for the male and female twins form natural pairs and are not independent.
      b. Independent samples would make more sense. The question of interest is not about any naturally occurring pairs.

11.15   Answers will differ for each student, but a possible answer is given for each part.
      a. The two populations are women who got married in 1979 and women who got married in 2009. The response variable is the woman's age when she got married. The question of interest is whether the average age at marriage for women changed between 1979 and 2009.
      b. The two populations are women with a college education and women without one. The response variable is the woman's age when she got married. The question of interest is whether the average age at marriage is the same for women with and without a college education.
      c. The two treatments are two different training programs for sales people in a company. The question of interest is which one is more effective in increasing average sales.
      d. A large company wants to know if it is more efficient to encourage customers to phone or send email when they have a request. The company measures a random sample of each type of request and measures how much time its employees spend dealing with the request.

11.17 s.e. $(\bar{x}) = \dfrac{s}{\sqrt{n}} = \dfrac{2}{\sqrt{64}} = 0.25$ cm. Roughly, over all possible samples of $n = 64$ from this population, the average difference between the sample mean and the population mean is about 0.25 cm.

11.19 This is paired data, so calculate the difference in pulse rates for each of the 50 people. Let s_d = standard deviation of the $n = 50$ *differences*.

Calculate the desired standard error of the mean as $\dfrac{s_d}{\sqrt{n}} = \dfrac{s_d}{\sqrt{50}}$.

11.21 Take a large enough sample or samples. One way to do this is to guess at what the standard deviation(s) will be and then figure out what size sample(s) are needed to produce a desired width for the confidence interval. For bell-shaped data, the standard deviation can be estimated if an approximate range for the type of measurement under investigation is known. Remember that for very large samples the range covers about six standard deviations, three on either side of the mean.

11.23 **a.** The value of t*increases when the confidence level is increased.
b. The degrees of freedom increase so the value of t* decreases.
c. When the degrees of freedom are increased the value of t* decreases.
d. When the degrees of freedom are essentially infinite the value of t* is found from the standard normal distribution.
e. When the standard deviation is decreased it has no effect on the value of t*.

11.25 **a.** $76 \pm 2.31 \dfrac{6}{\sqrt{9}}$ or 71.38 to 80.62; df $= n - 1 = 9 - 1 = 8$ for t^*.

b. $76 \pm 1.86 \dfrac{6}{\sqrt{9}}$ or 72.28 to 79.72

c. $76 \pm 1.75 \dfrac{6}{\sqrt{16}}$ or 73.375 to 78.625; df $= n - 1 = 16 - 1 = 15$ for t^*.

11.27 **a.** 2.52 (df $= 21$)
b. 2.13 (df $= 4$)
c. 4.60 (df $= 4$)

11.29 **a.** s.e. $(\bar{x}) = \dfrac{s}{\sqrt{n}} = \dfrac{6}{\sqrt{9}} = 2$. Roughly, this is the average difference between the sample and population means over all random samples of $n = 9$ that can be selected from this population.
b. $26.2 \pm (2.31 \times 2)$, or 21.6 to 30.8 sit-ups. (Using Table A.2, $t^* = 2.31$. Using software, $t^* = 2.306$.)
c. With 95% confidence, we can estimate that in the population of men represented by this sample the mean number of sit-ups in a minute is between 21.6 and 30.8.

11.31 Factors affecting width are (1) confidence level, (2) sample size, and (3) standard deviation.
Increasing the confidence level increases the width of an interval.
Increasing the sample size decreases the width of the interval because the degrees of freedom are larger, and thus the multiplier is smaller, and because the standard error is smaller.
The larger the standard deviation the greater the width of an interval.

11.33 The confidence interval is about 7.04 to 9.16 days, computed as $8.1 \pm 2.82 \times \dfrac{1.8}{\sqrt{23}}$.

Compared to part (b) of the previous exercise, only the multiplier t^* changes. Look in the .99 column of Table A.2 (df $= 24$) to locate t^*.

11.35 For this answer, we'll round the mean and standard deviation to the nearest dollar.
About 95% had textbook expenses in the interval $285 \pm (2)(\$96)$, which is $93 to $477
The calculation was $\bar{x} \pm 2 \times s$.

11.37 **a.** This is not a correct interpretation. The interval estimates the mean but does not provide information about the range of individual values.
b. This is a correct interpretation. We can be fairly confident that the mean is between 1.76 and 2.42 hours, so we can be fairly confident that it is under 3 hours.
c. This is not a correct interpretation. Any individual confidence interval is either correct or not. The confidence level addresses the long-run probability that the method will work. (See the discussion on page 374 in Chapter 10.)

11.39 $\text{s.e.}(\bar{x}) = \dfrac{s}{\sqrt{n}} = \dfrac{2.7}{\sqrt{81}} = 0.3$ inch. Approximate 95% confidence interval is

Sample estimate $\pm 2 \times$ *Standard error*, which is $64.2 \pm (2 \times 0.3)$, or 63.6 to 64.8 inches.

11.41 $\text{s.e.}(\bar{x}) = \dfrac{s}{\sqrt{n}} = \dfrac{2}{\sqrt{64}} = 0.25$ cm. The interval is 27 to 28 cm, computed as $27.5 \pm 2 \times 0.25$.

Calculate the interval as Sample estimate $\pm 2 \times$ Standard error, which here is $\bar{x} \pm 2 \times \text{s.e.}(\bar{x})$.

Interpretation: With approximate 95% confidence, we can say that in the population of men represented by this sample the mean foot length is between 27 and 28 cm.

11.43 Using the value of 2 as an approximation for the t* multiplier will work well only if the t* multiplier for 95% confidence is in fact close to 2. Examination of the values in Table A.2 will reveal that this is the case for degrees of freedom higher than about 30 or 40, but not for smaller degrees of freedom. For instance, for 5 degrees of freedom the correct multiplier is t* = 2.57, which is much larger than 2. The relationship df = $n - 1$ means that when n is very small, df will also be very small.

11.45 **a.** Using the information in Example 11.9 for a 90% confidence interval, we need only change the multiplier to find a 95% confidence interval. The multiplier from Table A.2 with df = 24 and confidence level = .95 is t* = 2.06. The 95% confidence interval is $5.36 \pm (2.06)(3.05)$, which is 5.36 ± 6.28 or -0.92 to 11.64.
b. No. The interval covers 0, and any value in the interval is a plausible population mean difference.
c. Two different variables are measured for each individual.

11.47 No. Any value in the interval is a plausible value for the difference in means. So the only way we could conclude that the two population means are identical is if the interval had no width, and 0 was the only value in it.

11.49 **a.** Compute interval as Sample estimate $\pm$ Multiplier $\times$ Standard error, which here is $\bar{d} \pm t^* \dfrac{s_d}{\sqrt{n}}$.

Sample estimate is $\bar{d} = 5$ ounces. Standard error is $\text{s.e.}(\bar{d}) = \dfrac{s_d}{\sqrt{n}} = \dfrac{7}{\sqrt{40}} = 1.107$.

Multiplier is t* = 2.02; in Table A.2, df = $n - 1 = 40 - 1 = 39$, which is not in the Table, so use 40 df and the .95 column.
The interval is $5 \pm (2.02 \times 1.107)$, or 5 ± 2.24, or 2.76 to 7.24 ounces.
b. Yes, it would be appropriate. The sample size of 40 is large enough. The interval is $5 \pm (2 \times 1.107)$, or 5 ± 2.214, or 2.786 to 7.214 ounces, which is very similar to the more precise interval in part (a).

11.51 **a.** Conclude that there is a difference between the population means. The value 0 would occur if there was no difference between population means. The interval encompasses the plausible values for the population mean difference. So if the interval does not cover 0 it says that a mean difference of 0 is not plausible.
b. We cannot conclude that there is a difference between the population means. The value 0 would occur if there was no difference between population means. So, if the interval covers 0 we cannot reject a statement (hypothesis) of no difference. We can only conclude that the true difference is likely to be somewhere in the range covered by the interval.

11.53 **a.** Here, Sample estimate $\pm$ Multiplier $\times$ Standard error is $\bar{x}_1 - \bar{x}_2 \pm t^* \times \text{s.e.}(\bar{x}_1 - \bar{x}_2)$

$\bar{x}_1 - \bar{x}_2 = 11.6 - 10.7 = 0.9$

$\text{s.e.}(\bar{x}_1 - \bar{x}_2) = \sqrt{\dfrac{s_1^2}{n_1} + \dfrac{s_2^2}{n_2}} = \sqrt{\dfrac{3.39^2}{16} + \dfrac{2.59^2}{10}} = 1.18$ (unpooled)

Multiplier is $t^* = 2.07$ (Use Table A.2 with df = 22)
b. *Interpretation:* With 95% confidence, we can say that in the populations of medical doctors and professors, the difference in mean testosterone levels is between −1.54 and 3.34. We are assuming that the samples are representative of all male medical doctors and university professors and that neither sample showed outliers or extreme skewness.
c. We cannot conclude that there is a difference between the mean testosterone levels of medical doctors and professors because the confidence interval for the difference in population means covers 0. All we can conclude is that anything in the interval is a plausible difference in the two population means.

11.55 **a.** $\bar{x}_1 - \bar{x}_2 = 57.5 - 55.3 = 2.2$ cm.

b. $\text{s.e.}(\bar{x}_1 - \bar{x}_2) = \sqrt{\dfrac{s_1^2}{n_1} + \dfrac{s_2^2}{n_2}} = \sqrt{\dfrac{2.4^2}{36} + \dfrac{1.8^2}{36}} = 0.5$ cm.

c. Approximate 95% confidence interval for the difference in population means is
Sample estimate $\pm 2\times$ *Standard error*, which is $2.2 \pm (2 \times 0.5)$, or 1.2 to 3.2 cm.

11.57 The interval is 2.86 to 3.14 cm, computed as $3.5 \pm 2 \times 0.32$ cm.
Parameter is $\mu_1 - \mu_2$ = difference between mean population foot lengths of men and women.
Calculate interval as $\bar{x}_1 - \bar{x}_2 \pm 2 \times \text{s.e.}(\bar{x}_1 - \bar{x}_2)$.

$\bar{x}_1 - \bar{x}_2 = 27.5 - 24 = 3.5$ cm.

Standard error is $\text{s.e.}(\bar{x}_1 - \bar{x}_2) = \sqrt{\dfrac{s_1^2}{n_1} + \dfrac{s_2^2}{n_2}} = \sqrt{\dfrac{2^2}{64} + \dfrac{2^2}{100}} = \sqrt{.0625 + .04} = 0.32$ cm.

Interpretation: With 95% confidence, we can say that in the population(s) represented by these sample(s), the difference in mean foot lengths of men and women is between 2.86cm and 3.14cm.

11.59 The variances (or, equivalently, the standard deviations) in the two populations are assumed to have the same value.

11.61 Confidence interval is about 2.62 to 4.58 days, computed as $3.6 \pm (2.01)(0.49)$.
Parameter is $\mu_1 - \mu_2$ = difference between mean days of symptoms in population of cold sufferers if taking placebo versus zinc lozenges.
Compute interval as Sample estimate $\pm$ Multiplier $\times$ Standard error:
Sample estimate is $\bar{x}_1 - \bar{x}_2 = 8.1 - 4.5 = 3.6$ days

Standard error (pooled) is $\text{s.e.}(\bar{x}_1 - \bar{x}_2) = \sqrt{\dfrac{s_p^2}{n_2} + \dfrac{s_p^2}{n_1}} = \sqrt{\dfrac{1.7^2}{23} + \dfrac{1.7^2}{25}} = 0.49$ days where $s_p = \sqrt{\dfrac{(n_1 - 1)s_1^2 + (n_2 - 1)s_2^2}{n_2 + n_2 - 2}} = \sqrt{\dfrac{(25 - 1) \times 1.6^2 + (23 - 1) \times 1.8^2}{23 + 25 - 2}} = 1.7$ days.

Multiplier is $t^* \approx 2.01$.

For pooled procedure, df $= n_1 + n_2 - 2 = 23 + 25 - 2 = 46$. Table A.2 does not have an entry for df $= 46$. We've used the entry for df $= 50$, although a more conservative procedure is to use the entry for df $= 40$. Minitab, Excel, or a calculator with the relevant capability could be used to determine that $t^* = 2.013$ for df $= 46$.

11.63 Formula for the 95% confidence interval is $\bar{x} \pm t^* \dfrac{s}{\sqrt{n}}$.

Calculation is $72.84 \pm 1.97 \times \dfrac{72}{\sqrt{205}}$, which is 72.84 ± 9.91 and this gives the desired result. Note: If the multiplier is found from Table A.2, the conservative choice is for df $= 100$, and $t^* = 1.98$.

11.65 The two confidence intervals given do not cover 1.0, the value of a relative risk that occurs if the risks under two conditions are the same. So, it is reasonable to conclude that in the population(s) represented by the sample(s), the risk of death during a twelve-year period is higher for people with abnormal heart rate recovery after treadmill exercise than it is for people with normal heart rate recovery.

11.67 **a.** Approximate 95% confidence interval is -0.08 to 0.12, computed as $0.02 \pm (2)(.050)$.

Parameter is $\mu_1 - \mu_2$ = difference in mean baseline symptoms scores for the populations represented by the placebo and calcium groups at the start of the study.

Sample estimate is $\bar{x}_1 - \bar{x}_2 = 0.92 - 0.90 = 0.02$

Standard error is $s.e.(\bar{x}_1 - \bar{x}_2) = \sqrt{\dfrac{s_1^2}{n_1} + \dfrac{s_2^2}{n_2}} = \sqrt{\dfrac{0.55^2}{235} + \dfrac{0.52^2}{231}} = 0.05$ (unpooled)

Multiplier ≈ 2 (samples are large)

b. The researchers would want this interval to cover 0. If the interval did not cover 0, it would show that the means of the placebo and calcium groups were different at the beginning of the study. The researchers want the two groups to be similar at the beginning of the study so they can say that any difference observed at the end of the study was caused by the difference in treatments (placebo or calcium).

11.69 **a.** Answer will vary.
b. Answer will vary. In most instances it will be between 91% and 99%.
c. Answer will vary. Expected percent is about 95%, so expected number is $150(.95) \approx 142$ or so. However, the actual number is a binomial random variable, with $p = .95$ and $n = 150$.

11.71 Specific intervals will vary. In all instances, the width of the interval will increase as the confidence level is increased.

11.73 **a.** The interval will be wider for 95% than for 90% confidence.
b. The interval will be more narrow if the sample size is doubled because the standard error will decrease.
c. Assuming the standard deviation stays the same, the width will remain the same if the sample size is the same. The observed value of a sample mean does not affect the width of the confidence interval.

11.75 **a.** A 99% confidence interval is about 1590.5 to 1613.5 mm.

Parameter is μ = mean height in the population. Confidence interval formula is $\bar{x} \pm t^* s.e.(\bar{x})$.

$\bar{x} = 1602$ mm, $s.e.(\bar{x}) = \dfrac{s}{\sqrt{n}} = \dfrac{62.4}{\sqrt{199}} = 4.4234$, and $t^* \approx 2.60$ (df $= 199 - 1 = 198$).

In Excel, TINV(.01,198) gives $t^* = 2.60$. In Table A.2, use the entry for df $= 100$ as an approximation.

101

Check necessary conditions: The sample size is sufficiently large. Assume the sample represents a random sample from the population.

Interpretation: We are 99% confident that the mean height of all women represented by this sample is between 1590.5 and 1613.5 mm.

b. The 99% confidence interval is about 62.6 inches to 63.5 inches. To determine the answer, convert the two ends of the interval found in part (a) to inches. For the lower value, the conversion is $(0.03937)(1590.5) \approx 62.6$. For the upper value, the conversion is $(0.03937)(1613.5) \approx 63.5$.

11.77 **a.** The last weight given for each team is an outlier. It is the weight of the coxswain, who gives instructions about the rowing cadence but does not row. A coxswain's weight is much less than the weights of the rowers, leading to the two substantial outliers. Because $n = 18$, the dataset doesn't qualify for either of the two situations on page 427.

b. A 90% confidence interval is about 183.7 to 199.6 pounds, computed as $191.63 \pm 1.89 \dfrac{11.84}{\sqrt{8}}$.

Parameter is μ = mean weight of population of all Cambridge crew team rowers over the years.

Formula is $\bar{x} \pm t^* \dfrac{s}{\sqrt{n}}$. Here, $t^* = 1.89$ (df = 8 – 1=7, use Table A.2). The Minitab output for this exercise gives other necessary statistics.

Minitab Output for Exercise 11.77b					
Variable	N	Mean	StDev	SE Mean	90.0% CI
Cambridge	8	191.63	11.84	4.19	(183.69, 199.56)

Check necessary conditions: A dotplot or a boxplot can be used to see that the weights of the eight rowers do not exhibit extreme skewness nor do they include outliers, so the necessary conditions are met. We assume that the sample represents a random sample of all Cambridge rowers over the years.

c. The 90% confidence interval is about –7.50 to 12.25 pounds if the unpooled standard error is used. With a pooled standard error, the interval is about –7.45 to 12.20 pounds. Outputs 1 and 2 for this part show results for the unpooled and pooled procedures, respectively.

Output 1 for Exercise 11.77c (Unpooled standard error)				
	N	Mean	StDev	SE Mean
Cambridg	8	191.6	11.8	4.2
Oxford	8	189.3	10.4	3.7
Difference = mu Cambridge - mu Oxford				
Estimate for difference: 2.38				
90% CI for difference: (-7.50, 12.25)				
DF = 13				

Output 2 for Exercise 11.77c (Pooled standard error)				
	N	Mean	StDev	SE Mean
Cambridg	8	191.6	11.8	4.2
Oxford	8	189.3	10.4	3.7
Difference = mu Cambridge - mu Oxford				
Estimate for difference: 2.38				
90% CI for difference: (-7.45, 12.20)				
Pooled StDev = 11.2				

Check necessary conditions: A comparative dotplot or boxplot can be used to see that the weights of the rowers at each school do not exhibit extreme skewness and there are no outliers, so the necessary conditions are met. We assume that the samples represent larger populations of men who have rowed for these schools over the years.

The question asked for a justification of the use of either the pooled or unpooled version. In this example, either method is acceptable because the standard deviations for the two groups are very similar. In fact, when sample sizes are equal, the pooled and unpooled standard errors are exactly the same. The only difference in the two methods is in the degrees of freedom used to find the multiplier. For the unpooled version df=13 (from Minitab) while for the pooled version df = 8 + 8 − 2 = 14, leading to multipliers of 1.77 and 1.76, respectively. Therefore, the two methods are essentially the same for this example.

d. *Interpretation*: With 90% confidence, we can say that the difference between the mean weights in the two populations represented by the Cambridge and Oxford rowers is between −7.5 and 12.2 pounds. Based on this interval, we cannot determine which population of rowers, if either, has the greater mean weight.

11.79   **a.** It is important to recognize that the appropriate population depends on what was measured, which in this case is TV viewing hours. The TV viewing habits of the students in a statistics class at a particular university at a particular time may represent the TV viewing habits of all students at that school at that time, or the subset who will take a statistics class, or perhaps students at all similar schools who take statistics at some time in college. So the Fundamental Rule for Using Data for Inference holds in this case if the population is defined appropriately. The results can be extended only to whatever population we think the students in the class represent for the question of TV viewing hours.

b. No. The interval tells us that the *average* of TV viewing hours for all students in the population is between 1.842 and 2.338 hours. It doesn't tell us anything about individual students.

11.81   **a.** A 95% confidence interval is about 1.60 to 2.88 years, computed as $2.24 \pm (2)(0.3145)$

Parameter is μ_d = mean difference in ages of husband and wife in population of British married couples.

For computing the interval is Sample estimate ± Multiplier × Standard error:

Sample estimate is $\bar{d} = 2.24$ years.   Standard error is $\text{s.e.}(\bar{d}) = \dfrac{s_d}{\sqrt{n}} = \dfrac{4.1}{\sqrt{170}} = 0.3145$

Multiplier ≈ 2 (because the sample size is large).

Check necessary conditions: The sample size is sufficiently large. We assume the sample represents a random sample of British couples.

Interpretation: With 95% confidence, we can say that in the population of British married couples, the mean difference (husband age – wife age) is between 1.60 years and 2.88 years.

b. This is paired data. The differences between the ages of the husband and wife can be determined for each couple and are not independent.

11.83   **a.** Observed difference = 2.37−1.95 = 0.42 hours. This is a statistic.

b. Females: Standard error of mean = $\dfrac{1.51}{\sqrt{116}} = 0.14$.

Males: Standard error of mean = $\dfrac{1.87}{\sqrt{59}} = 0.24$.

c. We cannot conclude that there is a difference in the population represented by the sample. The confidence interval for the difference in population means contains the value 0, so we cannot reject the possibility that the population difference is 0. With 95% confidence we can conclude that the actual difference in population means is between −0.14 hour (about 8 minutes) more viewing for females and +0.97 hour (about 58 minutes) more viewing for males.

d. The formula is $\bar{x}_1 - \bar{x}_2 \pm 2 \times \sqrt{\dfrac{s_1^2}{n_1} + \dfrac{s_2^2}{n_2}}$.

Substitution of relevant values gives $(2.37 - 1.95) \pm 2 \times \sqrt{\dfrac{1.87^2}{59} + \dfrac{1.51^2}{116}}$.

Note: Assignment of group numbers (1 or 2) to the specific groups is essentially arbitrary. We've designated the men as group 1 so that the difference in part (a) is positive.

11.85   **a.** A "success" for a researcher is that the computed interval covers the population mean.

b. The probability of a success is $p = .90$. This probability is the relative frequency of times over all possible random samples from the population that an interval covers the population mean

c. Expected number of intervals covering the population mean is $np = (100)(.90) = 90$.

d. No, each researcher will not know whether he or she has a "successful" interval. That would require knowing the value of the population mean.

e. Probability = $(.90)^{100}$ = .000027 that all intervals cover the population mean. Use the extension of Rule 3b on p.233. It can also be calculated using Minitab or Excel to calculate $P(X = 100)$ in a binomial distribution with $n = 100$ and $p = .9$. For instance the Excel command BINOMDIST(100,100,0.90,FALSE) could be used.

11.87 **a.** Parameter is μ_1 = mean right handspan in population of men represented by the sample.

Confidence Interval: Minitab gives the 95% confidence interval for μ_1 as 22.247 to 22.868 cm.

Output for Exercise 11.87a					
Variable	N	Mean	StDev	SE Mean	95.0% CI
RtSpan_M	87	22.557	1.458	0.156	(22.247, 22.868)

Interpretation: With 95% confidence, we can say that in the population of men represented by the sample, the mean stretched right handspan is between 22.247 and 22.868 cm.

Necessary assumptions: Assume that with regard to stretched right handspans this sample of college men and women represents a random sample from a larger population of men and women.

Check necessary conditions: The sample size is sufficiently large, although even with a large sample it is useful to examine plots of the data to make sure there are no extreme outliers. Here, a boxplot shows that the distribution is roughly symmetric and there is only one mild, probably harmless outlier.

b. Parameter is μ_2 = mean right handspan in population of women represented by the sample.

Confidence Interval: Minitab gives the 95% confidence interval for μ_2 as 19.672 to 20.362 cm.

Output for Exercise 11.87b					
Variable	N	Mean	StDev	SE Mean	95.0% CI
RtSpan_F	103	20.017	1.764	0.174	(19.672, 20.362)

Interpretation: With 95% confidence, we can say that in the population of women represented by the sample, the mean stretched right handspan is between 19.672 and 20.362 cm.

Necessary assumptions: Stated in part (a).

Check necessary conditions: The sample size is sufficiently large, although a plot shows that there are two notable outliers. If these outliers are removed, the 95% confidence interval is 19.875 to 20.447 cm, so they don't have a great influence on the result.

c. Parameter is $\mu_1 - \mu_2$. Confidence interval (unpooled version) given by Minitab is 2.079 to 3.002 cm. Pooled version is 2.072 to 3.009 cm. If the two female outliers are omitted, the confidence interval is about 1.98 to 2.82 cm. All of these intervals are roughly 2 to 3 cm.

Interpretation: With 95% confidence, we can say that of men and women in the population(s) represented by the sample(s), the difference in the mean stretched handspans is between approximately 2 and 3 cm.

Assumptions and conditions: Given in parts (a) and (b).

11.89 **a.** The 98% confidence interval given by Minitab is 183.11 to 203.16.

	Output for Exercise 11.89a				
Variable	N	Mean	StDev	SE Mean	98.0% CI
control	30	193.13	22.30	4.07	(183.11, 203.16)

Interpretation: With 98% confidence, we can say that in the population of individuals who have not had a heart attack, the mean cholesterol level is between 183.11 and 203.16.

Assumption and necessary conditions: The sample represents a random sample from a larger population of individuals who have not had a heart attack. The data are roughly bell-shaped and there are no outliers. A comparative dotplot is shown here because part (b) asks about the patients who have had a heart attack.

Figure for Exercise 11.89 parts a and b

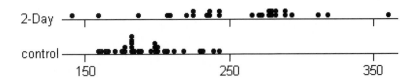

b. The 98% confidence interval given by Minitab is 231.63 to 276.22.

Output for Exercise 11.89b					
Variable	N	Mean	StDev	SE Mean	98.0% CI
2-Day	28	253.93	47.71	9.02	(231.63, 276.22)

Interpretation: With 98% confidence, we can say that in the population of individuals who have had a heart attack, the mean cholesterol level two days after the attack is between 231.63 and 276.22.

Assumption and necessary conditions: The sample represents a random sample from a larger population of individuals who have had a heart attack. The data are roughly bell-shaped and there are no outliers.

c. A 98% confidence interval (unpooled procedure) for the difference in population means is given by Minitab as 36.74 to 84.85.

```
                Output for Exercise 11.89c
Two-sample T for 2-Day vs control
          N      Mean      StDev    SE Mean
2-Day    28     253.9      47.7       9.0
control  30     193.1      22.3       4.1

Difference = mu 2-Day - mu control
Estimate for difference:  60.80
98% CI for difference: (36.74, 84.85)
.......  DF = 37
```

Interpretation: With 98% confidence we can say that in the population of people who have suffered a heart attack, the mean cholesterol (measured two days after the attack) is between 36.74 and 84.85 points higher than the mean cholesterol in a population of people who have not had a heart attack.

Assumption and necessary conditions: See parts (a) and (b).

CHAPTER 12
ODD-NUMBERED SOLUTIONS

12.1 **a.** All babies.
b. p = proportion of babies in the population born during the 24-hour period surrounding a full moon.
c. H_0: $p = 1/29.53$

12.3 **a.** The populations of interest are all 21-year-old men and all 21-year-old women.
b. Researchers are interested in the hypothesis that the proportions of 21-year-old men and women who have high school diplomas are the same.

12.5 **a.** Null hypothesis because it is a statement of "no difference."
b. Alternative hypothesis because it states that there is a difference.
c. Alternative hypothesis because it states that there is a difference.
d. Null hypothesis. Even though the hypothesis includes values that indicate a difference, it also includes the statement that the proportions are equal.

12.7 **a.** No. A null hypothesis is a statement about a population value, not a sample value as in the statement for this exercise.
b. Yes. It's a statement of no difference and could be interpreted to apply to all newborns.
c. No. This statement would more appropriately be an alternative hypothesis (a statement of a difference).

12.9 **a.** H_0: Increasing speed limits on interstate highways does not increase the fatality rate.
   H_a: Increasing speed limits on interstate highways increases the fatality rate.
The hypothesis test is one-sided.
The alternative hypothesis specifies a single direction, a positive association between speed limits and fatality rates.
b. H_0: Probability = .5 that a coin spun on its edge lands heads up.
   H_a: Probability ≠ .5 that a coin spun on its edge lands heads up.
The hypothesis test is two-sided. The alternative hypothesis does not give a specific direction for how the probability might differ from .5.

12.11 **a.** There are 5 choices so $p = 1/5 = .2$.   Random selection means that each of the five choices is equally likely to be chosen.
b. H_0: $p = .2$ (random selection)
   H_a: $p < .2$ (less often than random selection would give)

12.13 The decision to use a one-sided or two-sided alternative hypothesis should be made before looking at the sample data. The alternative hypothesis usually expresses the research hypothesis, which in turn gives the reason for collecting the data. The reason for collecting the data should be defined before the data are collected

12.15 **a.** H_0: $p = .06$ (percent is still 6%)
   H_a: $p > .06$ (percent now greater than 6%)
The null hypothesis could also be written as H_0: $p \leq .06$. It's possible the percent has decreased.
In words:
Null: The percent of children in the state who live with their grandparent(s) remains at 6% (or possibly even less).
Alternative: The percent of children in the state who live with their grandparent(s) has increased since the last census and is now greater than 6%.
b. Population is all school-aged children living in the state at the time of the study.
Proportion of interest = proportion of them who live with one or more grandparents.

12.17   **a.** $H_0: \mu_1 - \mu_2 = 0$ where μ_1 is the mean weight of all newborn babies in England and μ_2 is the mean weight of all newborn babies in the United States.
   b. The null value is 0 because when the two mean weights are the same, $\mu_1 - \mu_2 = 0$.

12.19   **a.** Cannot reject the null hypothesis. The p-value is not less than the level of significance.
   b. Reject the null hypothesis. The p-value is less than the level of significance.
   c. Cannot reject the null hypothesis. The p-value is less than the level of significance.
   d. Reject the null hypothesis. The p-value is less than the level of significance.

12.21   **a.** There is not enough evidence to conclude that smokers are more likely to get the disease.
   b. There is enough evidence in the sample to conclude that smokers are move likely to get the disease in the population.

12.23   Yes. The critical value is the test statistic value for which the p-value equals the designated level of significance. More extreme values of the test statistic will have smaller p-values that are less than the level of significance.

12.25   **a.**   The parameter of interest is p = population proportion of artists who are left-handed. The null value is .10.
   b. $H_0: p = .10$ (same as proportion in general population)
       $H_a: p > .10$ (greater than proportion in general population)
   c. The sample proportion is $\hat{p} = 18/150 = .12$ that are left-handed.
   d. The p-value is the probability that the sample proportion would be .12 or larger (for a sample of $n = 150$) if the population proportion actually is .10.
   e. The test statistic is $z = \dfrac{\text{Sample estimate - Null value}}{\text{Null standard error}}$. The numerator is $(.12 - .10)$.

12.27   The p-value will be smaller for Researcher B because that researcher's sample proportion is farther from $p = .25$ (the null value) in the direction specified by the alternative hypothesis. So, it is stronger evidence against the null hypothesis. In general, the stronger the evidence is against a null hypothesis the smaller is the p-value. If the null hypothesis were true, the distribution of possible sample proportions would be approximately a normal curve centered at .25 (by the Rule for Sample Proportions, Chapter 9). The probability of observing a sample proportion of .33 or larger is smaller than the probability of observing a sample proportion of .29 or larger (so the p-value is smaller for Researcher B). The following figure illustrates the situation.

Figure for Exercise 12.27

p-values for A and B are probabilities to the right of the observed proportions

0.25   0.29   0.33
(null)   (A)   (B)

12.29   **a.** False. The p-value is the probability the test statistic would be as extreme as it is or more so, computed assuming the null is true.
   b. True by definition.

108

c. False. A type 2 error can occur only when the alternative is true.

d. False. Type 1 and 2 errors are not complementary (opposite) events. A type 1 error can be made only when the null hypothesis is true and a type 2 error can be made only when the alternative hypothesis is true.

12.31 **a.** The correct decision has been made. The null is true and it is not rejected.

b. A type 2 error has occurred. The alternative is true, but the null is not rejected.

c. A type 1 error has occurred. The null is true, but it has been rejected.

d. The correct decision has been made. The alternative is true and the decision is in favor of the alternative.

12.33 **a.** The company believes the proportion requiring a hospital stay will decrease, but in fact it will not decrease.

b. The company believes the proportion requiring a hospital stay will not decrease, but in fact it will.

c. Type 1. It will offer the coverage, but the proportion requiring a stay will not decrease. The company's profit is likely to decrease as a result due to the increased payments made for alternative medicine treatments. A type 2 error would not change the company's present situation as it would not offer the coverage, so the proportion requiring a stay would not change.

d. Type 2. More customers will require a hospital stay than necessary because the coverage for alternative medicine will not be offered (because the company believes it will not be helpful).

12.35 In this situation, a type 1 error would occur if the patient is incorrectly diagnosed as having the disease. A type 2 error would occur if the patient was incorrectly believed to not have the disease.

a. The answer will vary. An example: A type 1 error would be more serious if the diagnosis of that the disease is present leads to a major surgery. For example, in certain elderly men prostate cancer grows very slowly and may never become life threatening, so a type 2 error may not be so serious. A type 1 error could lead to unnecessary surgery and would be more serious.

b. The answer will vary. An example: A type 2 error would be more serious for the diagnosis of an infection that could be cured by antibiotics but gets very serious if untreated (the consequence of a type 2 error). Although there is some concern about overusing antibiotics, it would not be particularly harmful for the patient to take them even though an infection is not present (the consequence of a type 1 error).

12.37 **a.** A type 1 error occurs if the politician believes that more than one-half of voters in his district support the new tax bill when the proportion really is not more than one-half. The consequence is that he would vote for a bill that is not supported by a majority of the voters in his district.

b. A type 2 error occurs if the politician believes the proportion of voters in the district supporting the tax bill is not a majority when really it is a majority. The consequence is that that he would not vote for a bill that is supported by a majority of the voters.

c. A type 1 error is probably more serious because this involves voting for the bill that implements substantial change, when a majority of the voters do not support it.

d. Type 1. The politician decided in favor of the alternative hypothesis, so the potential mistake is that he has incorrectly rejected the null hypothesis.

12.39 The power of any hypothesis tests done will be greater with the larger sample size.

12.41 **a.** p-value = .0358. This is the combined probability to the right of $z = 2.10$ and left of $z = -2.10$. Table A.1 can be used to find that $P(Z < -2.10) = .0179$. By symmetry, the area to the right of $z = 2.10$ is also .0179. So, the p-value = $2 \times .0179 = .0358$.

Figure for Exercise 12.41a

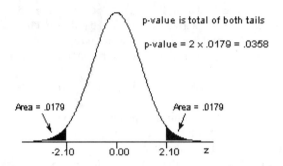

b. *p*-value = .0228. This is the probability (area) to the left of $z = -2.00$. Table A.1 can be used to find $P(Z < -2) = .0228$.

Figure for Exercise 12.41b

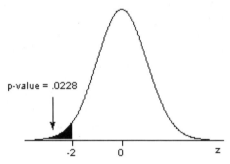

c. *p*-value = .1379. This is the probability (area) to the left of $z = -1.09$. Table A.1 can be used to find $P(Z < -1.09) = .1379$.

Figure for Exercise 12.41c

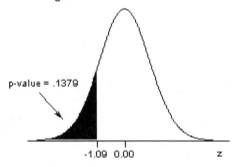

d. *p*-value ≈ .00001. This is the probability (area) to right of $z = 4.25$. By the symmetry of the normal curve, this probability equals the area to the left of $z = -4.25$. In Table A.1, use the "In the Extreme" section to estimate the probability. A probability (to the left) is given for $z = -4.26$, which is close enough for this exercise. Software or a calculator with the relevant capability could also be used to find this probability.

110

Figure for Exercise 12.41d

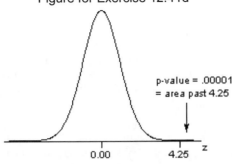

p-value = .00001
= area past 4.25

12.43 **a.** Yes. The sample is a random sample, and the sample size is large enough because $np_0 = (20)(.50) = 10$, and $n(1-p_0) = (20)(1-.50) = 10$.

b. No. The sample size is not large enough because $np_0 = (20)(.10) = 2$ is smaller than 10.

12.45 **a.** The proportion of all stockbrokers believing the market will go up next year.
b. The proportion of all mall visitors that buy something.

12.47 **a.** $z = \dfrac{\hat{p} - p_0}{\sqrt{\dfrac{p_0(1-p_0)}{n}}} = \dfrac{.3-.2}{\sqrt{\dfrac{.2(1-.2)}{500}}} = 5.59$.

b. $z = \dfrac{\hat{p} - p_0}{\sqrt{\dfrac{p_0(1-p_0)}{n}}} = \dfrac{.5-.8}{\sqrt{\dfrac{.8(1-.8)}{200}}} = -10.61$.

12.49 The null standard error uses the hypothesized null value p_0, while the standard error uses the observed sample proportion $\hat{p}$.

Null standard error = $\sqrt{\dfrac{p_0(1-p_0)}{n}}$ while standard error = $\sqrt{\dfrac{\hat{p}(1-\hat{p})}{n}}$

12.51 **a.** $H_0: p \le .50$ (one-half or less do)
$H_a: p > .50$ (more than one-half do)
p = proportion of all adult American Catholics who favor allowing women to be priests.
Usually the "yes" answer to the research question is the alternative hypothesis, and that is what has been done here.
b. The sample was randomly selected, which is one of the necessary conditions. The sample size is large enough, which is the other necessary condition. Both np_0 and $n(1-p_0)$ are greater than 10, as they should be to use a z-statistic. Here, $n = 507$ and $p_0 = .50$.

c. $z = \dfrac{\text{Sample estimate - Null value}}{\text{Null standard error}} = \dfrac{\hat{p} - p_0}{\sqrt{\dfrac{p_0(1-p_0)}{n}}} = \dfrac{.59-.50}{\sqrt{\dfrac{.5(1-.5)}{507}}} = \dfrac{.09}{.0222} = 4.05$

d. p-value $\approx .00003$. This is the probability (area) to the right of $z = 4.05$. To find the p-value exactly, use software or a calculator that can give normal curve probabilities. Table A.1 can be used to approximate the p-value. Due to symmetry, the area to the right of $z = 4.05$ equals the area to the left of $z = -4.05$. Near the bottom of left-side page of Table A.1, probabilities (areas) to the left of -3.72 and -4.25 are given as .0001 and .00001respectively. The correct p-value is between these two values. Typically, this information may be stated as p-value $< .0001$.

111

e. Reject the null hypothesis and decide in favor of the alternative hypothesis. The p-value is smaller than 0.05. This is evidence that more than one-half ($p > .50$) of all adult American Catholics favor allowing women to be priests.

f. Standard error $= s.e.(\hat{p}) = \sqrt{\dfrac{\hat{p}(1-\hat{p})}{n}} = \sqrt{\dfrac{.58(1-.58)}{507}} = .022$.

An approximate 95% confidence interval is $\hat{p} \pm 2 \times standard\ error$, which is $.59 \pm (2)(.022)$, or .55 to .63. All values in this confidence interval are greater than .50. This confirms the conclusion that a majority of adult American Catholics favor allowing women priests.

12.53   **a.** We cannot reject the null hypothesis because the p-value (.07) is greater than .05.
         b. The p-value for the one-sided test is $.07/2 = .035$. We would be able to decide in favor of the alternative hypothesis because this p-value is smaller than .05 (the usual criterion for statistical significance).

12.55   <u>Step 1</u>: $H_0: p \geq .5$ (one-half or more of the population of U.S. adults is "very happy")
              $H_a: p < .5$ (only a minority is "very happy")
$p=$ proportion of adults in the U.S. population who classify themselves as "very happy"
<u>Step 2</u>: The necessary conditions for using the z-statistic are present. The sample was randomly selected and the sample size is large enough so that both np_0 and $n(1-p_0)$ are greater than 10, as they should be to use a z-statistic. Here, $n = 1052$ and $p_0 = .5$.

The test statistic is $z = \dfrac{\text{Sample estimate - Null value}}{\text{Null standard error}} = \dfrac{\hat{p} - p_0}{\sqrt{\dfrac{p_0(1-p_0)}{n}}}$

Sample estimate $= \hat{p} = .47$

$z = \dfrac{\hat{p} - p_0}{\sqrt{\dfrac{p_0(1-p_0)}{n}}} = \dfrac{.47 - .5}{\sqrt{\dfrac{.5(1-.5)}{1052}}} = \dfrac{-.03}{.0154} = -1.95$

<u>Step 3</u>: p-value $= .0256$. This is the probability (area) to the left of $z = -1.95$, and it is illustrated in the figure below. Table A.1 can be used to find that $P(z < -1.95) = .0256$.

Figure for Exercise 12.55

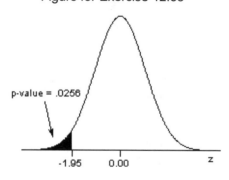

<u>Step 4</u>: Reject the null hypothesis. The result is statistically significant because the p-value is less than .05, the usual standard for significance.
<u>Step 5</u>: The data support the journalist's headline that a minority (less than .5) of all U.S. adults classify themselves as "very happy" but it should be noted that the observed proportion was only slightly less than .5. The result here has statistical significance, but it may not have practical significance. The headline could just as well have been "About one-half of U.S. adults are very happy."
Note: It can be argued that a two-sided test should be done here, in which case the null hypothesis would not be rejected. If the journalist looked at the data first and then made a decision about which kind of headline to write, it would be misleading to use a one-sided alternative. In that case, we can assume that if

the data had been in the other direction the headline would have ended with "in the majority" instead of "in the minority." However, the journalist is reported to be a pessimist, so we assume he or she was only interested in finding a significant result in this direction, and hence, used a one-sided alternative.

12.57 $P(X \geq 17)$ for a binomial random variable with $n = 50$ and $p = .25$. (See the table on page 472.)

12.59   **a.** $P(X \leq 17) = .0983$. Use a binomial distribution with $n = 50$ and $p = .25$.
b. We cannot reject the null hypothesis using $\alpha = .05$ (but we could using $\alpha = .10$).

12.61   **a.** p-value $= .0401$. It is the area (probability) to the right of $z = 1.75$ under a standard normal curve. $P(z > 1.75) = P(z \leq -1.75) = .0401$. Equivalently, $P(z > 1.75) = 1 - P(z \leq 1.75) = 1 - .9599$.

Figure for Exercise 12.61a

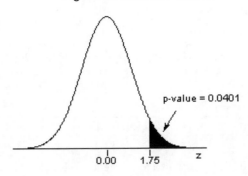

b. p-value $= .9599$. It is the area (probability) to the right of $z = -1.75$ under a standard normal curve. $P(z > -1.75) = P(z \leq 1.75) = .9599$. Equivalently, $P(z > -1.75) = 1 - P(z \leq -1.75) = 1 - .0401$.

Figure for Exercise 12.61b

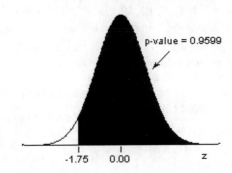

12.63   **a.** Reject the null hypothesis (accept the alternative); p-value $= P(z \leq -1.99) = .0233$ is less than .05.
b. Reject the null hypothesis (accept the alternative); p-value $= P(z > 1.78) = P(z \leq -1.78) = .0375$ is less than .05.

12.65   **a.** H_0: $p_1 - p_2 = 0$, or equivalently, $p_1 = p_2$ (no difference in proportions)
        H_a: $p_1 - p_2 < 0$, or equivalently, $p_1 < p_2$ (proportion lower for men)
b. $\hat{p}_1 = .4, \hat{p}_2 = .6$, found as 20/50 and 30/50, respectively.
c. $\hat{p} = \dfrac{20 + 30}{50 + 50} = \dfrac{50}{100} = .5$
d. Test statistic is $z = \dfrac{\text{Sample statistic - Null value}}{\text{Null standard error}} = \dfrac{(.4 - .6) - 0}{.10} = -2.00$ .where

113

Null standard error $= \sqrt{\dfrac{\hat{p}(1-\hat{p})}{n_1} + \dfrac{\hat{p}(1-\hat{p})}{n_2}} = \sqrt{\dfrac{.5(1-.5)}{50} + \dfrac{.5(1-.5)}{50}} = .10$

e. p-value $= P(z \le -2.00) = .0228$ (in Table A.1); reject the null hypothesis. Conclude that the proportion of men who write a shopping list is less than the proportion of women who do so.

12.67 <u>Step 1</u>: H_0: $p_1 - p_2 = 0$, or equivalently, $p_1 = p_2$ (no difference in proportions)

H_a: $p_1 - p_2 > 0$, or equivalently, $p_1 > p_2$ (proportion higher for women)

p_1 = proportion of women in the population who would claim they would return the money
p_2 = proportion of men in the population who would claim they would return the money
<u>Step 2</u>: There are two independent samples and the observed counts in both categories (yes and no) are greater than 10 for both males and females. We must assume the samples represent random samples from the population of all college students.

Test statistic is $z = \dfrac{\text{Sample statistic - Null value}}{\text{Null standard error}} = \dfrac{.1965 - 0}{.0602} = 3.26$. Details are:

Women, $\hat{p}_1 = \dfrac{84}{93} = .9032$; men, $\hat{p}_2 = \dfrac{53}{75} = .7067$; $\hat{p}_1 - \hat{p}_2 = .9032 - .7067 = .1965$

Combined $\hat{p} = \dfrac{84 + 53}{93 + 75} = .8155$

Null standard error $= \sqrt{\dfrac{\hat{p}(1-\hat{p})}{n_1} + \dfrac{\hat{p}(1-\hat{p})}{n_2}} = \sqrt{\dfrac{.8155(1-.8155)}{93} + \dfrac{.8155(1-.8155)}{75}} = .0602$

<u>Step 3</u>: p-value $= .0006$ (or $\approx .001$). It is the area to the right of 3.26 under a standard normal curve. $P(z > 3.26) = P(z \le -3.26) = .0006$ (in Table A.1).
<u>Steps 4 and 5</u>: We can reject the null hypothesis. The conclusion is that in the population(s) represented by the sample(s) a higher proportion of women than men would claim they would return the money if they found a wallet on the street.

12.69 <u>Step 1</u>: H_0: $p_1 - p_2 = 0$, or equivalently, $p_1 = p_2$ (no difference in proportions)

H_a: $p_1 - p_2 > 0$, or equivalently, $p_1 > p_2$ (proportion higher for women)

p_1 = proportion of women in the population who would say "more strict"
p_2 = proportion of men in the population who would say "more strict"
<u>Step 2</u>: There are two independent samples assumed to be randomly selected, and the observed counts in both categories (more strict or other answer) are greater than 10 for both groups (women and men).

Test statistic is $z = \dfrac{\text{Sample statistic - Null value}}{\text{Null standard error}} = \dfrac{.20 - 0}{.03193} = 6.26$. Details are:

Sample statistic $= \hat{p}_1 - \hat{p}_2 = .72 - .52 = .20$

Combined $\hat{p} = \dfrac{n_1 \hat{p}_1 + n_2 \hat{p}_2}{n_1 + n_2} = \dfrac{538(.72) + 493(.52)}{538 + 493} = .6244$

Null standard error $= \sqrt{\dfrac{\hat{p}(1-\hat{p})}{n_1} + \dfrac{\hat{p}(1-\hat{p})}{n_2}} = \sqrt{\dfrac{.6244(1-.6244)}{538} + \dfrac{.6244(1-.6244)}{493}} = .03193$

<u>Step 3</u>: p-value $< .000000001$. It is the area to the right of 6.26 under a standard normal curve. $P(z > 6.26) = P(z \le -6.26)$. At the bottom of the left page in Table A.1, a cumulative probability is given for $z = -6.00$. The p-value must be less than that probability because -6.26 is more extreme than -6.00.
<u>Steps 4 and 5</u>: We can reject the null hypothesis. The conclusion is that in the population(s) represented by the sample(s) the proportion of women who would say there should be "more strict" laws covering the sale of firearms is higher than the proportion of men who would say this.

12.71 <u>Step 1</u>: H_0: $p_1 - p_2 = 0$, or equivalently, $p_1 = p_2$ (no difference in proportions)

H_a: $p_1 - p_2 > 0$, or equivalently, $p_1 > p_2$ (proportion higher for women)

p_1 = proportion of women in the population who would say yes

114

p_2 = proportion of men in the population who would say yes

Step 2: There are two independent samples and the observed counts in both categories (yes or no) are greater than 10 for both groups (women and men). We must assume the samples represent random samples from the population of all college students.

Test statistic is $z = \dfrac{\text{Sample statistic - Null value}}{\text{Null standard error}} = \dfrac{.185 - 0}{.07708} = 2.40$. Details are:

Sample statistic = $\hat{p}_1 - \hat{p}_2 = .611 - .426 = .185$

Combined $\hat{p} = \dfrac{n_1\hat{p}_1 + n_2\hat{p}_2}{n_1 + n_2} = \dfrac{131(.611) + 61(.426)}{131 + 61} = .552$

Null standard error = $\sqrt{\hat{p}(1-\hat{p})(\dfrac{1}{n_1} + \dfrac{1}{n_2})} = \sqrt{.552(1-.552)(\dfrac{1}{131} + \dfrac{1}{61})} = 0.07708$

Step 3: p-value = .0082. It is the area to the right of $z = 2.40$ under a standard normal curve. $P(z > 2.40) = P(z \le -2.40) = .0082$. Equivalently, $P(z > 2.40) = 1 - P(z \le 2.40) = 1 - .9918 = .0082$.
Steps 4 and 5: We can reject the null hypothesis. The conclusion is that in the population(s) represented by the sample(s) a higher proportion of women than men would claim they would date someone with a great personality even if they did not find them physically attractive.

12.73 **a.** H_0: $p_1 - p_2 = 0$, or equivalently, $p_1 = p_2$
  H_a: $p_1 - p_2 > 0$, or equivalently, $p_1 > p_2$
  p_1 = proportion ever bullied in the population of short students, and
  p_2 = proportion ever bullied in the population of students who are not short
  b. Step 2: Sample sizes are sufficiently large so that observed counts in both categories (bullied or not) are greater than 10 for both groups.
  Test statistic is $z = 3.02$.
  Steps 3, 4, and 5: p-value = .001. Reject the null hypothesis. Conclusion is that the proportion ever bullied is higher in the population of short students.
  c. The upper limit is 1, the largest possible (positive) difference between two proportions. The complete interval is .09192 to 1.
  d. The confidence interval falls entirely in the alternative hypothesis region, and does not cover 0, so it is evidence in favor of the alternative hypothesis.

12.75 **a.** $\dfrac{.75 - .70}{.0725} = 0.690$; $p = 2 \times P(Z < -0.69) = (2)(.2451) = .4902$, so fail to reject null hypothesis.

  b. $\dfrac{.75 - .70}{.0458} = 1.091$; $p \approx 2 \times P(Z < -1.09) = (2)(.1379) = .2758$, so fail to reject null hypothesis.

  c. $\dfrac{.75 - .70}{.0205} = 2.439$; $p \approx 2 \times P(Z < -2.44) = (2)(.0073) = .0146$, so reject the null hypothesis.

  d. $\dfrac{.75 - .70}{.0145} = 3.448$ $p \approx 2 \times P(Z < -3.45) = (2)(.0003) = .0006$, so reject the null hypothesis.

Overall comment on sample size: With the same values for $\hat{p}$ and p_0, a larger sample size gives a greater value for the z-statistic and a smaller value for the p-value.

12.77 **a.** Assuming the true proportion actually is higher in the second group, $n = 1000$. The larger sample is more likely to provide a statistically significant result.
  b. $p = .45$. The farther the true proportion is from the null value of .20 (as long as it's in the direction stated in the alternative hypothesis), the more likely it is that the sample p will be significantly greater than .20.
  c. $\hat{p} = .45$. The value of the z-statistic will be greater because the difference between .45 and .20 (the null value) is greater than the difference between .25 and .20.

115

12.79 *p*-value. For a non-significant result, the value gives information about whether the sample result was close to being significant or not. For a significant result, the value gives information about the strength of the result.

12.81 Probably the finding was that there was no *significant* difference, in the statistical meaning, not the ordinary language meaning. This finding could have occurred because there really is no difference, or because the sample was too small to detect it and a type 2 error was made.

12.83 **a.** Yes, the difference is statistically significant. The reported *p*-value (.005) is less than .05, the usual standard for significance.
b. No, this is not a contradiction. There is not much (if any) practical importance to the observed difference in incidence of drowsiness (6% versus 8%), but the large sample sizes led to a *statistically* significant difference.
c. A statistically significant difference indicates that the difference in the population is not zero but does not indicate that it has any practical significance. The meaning should be clarified when the word is used.

12.85 **a.** .3776 (read directly from the given output)
b. With a sample of $n = 100$, the probability is .5740 that the null hypothesis will be rejected if the true probability of a correct guess is $p = .33$.
c. Yes, the output shows that the power is .9705 for $n = 400$, so the researcher will have greater than .95 probability of detecting ESP when true p for success is .33.
d. Higher. For a specific sample size, the greater the difference between the true p and the null value, the greater the power of the hypothesis test.

12.87 Steps 3 and 4 are different. For Step 3, instead of finding the *p*-value, we find the rejection region. For Step 4, instead of using the *p*-value to make a decision we determine whether the z test statistic falls in the rejection region.

12.89 You would need to know how large the difference in weight loss was for the two groups. If the difference in weight loss is very small (but not 0) it could be statistically significant, but not have much practical importance.

12.91 Step 1: $H_0: p = .1$ (proportion left-handed same as in national population)
$\qquad$ $H_a: p \neq .1$ (proportion left-handed not same as in national population.
Step 2: The necessary conditions for using the z-statistic are present. The sample is assumed to be representative of the larger population and the sample size is large enough so that both np_0 and $n(1 - p_0)$ are greater than 10. Here, $n = 240$ and $p_0 = .1$.

The test statistic is $z = \dfrac{\text{Sample estimate - Null value}}{\text{Null standard error}} = \dfrac{\hat{p} - p_0}{\sqrt{\dfrac{p_0(1 - p_0)}{n}}}$.

Sample estimate $= \hat{p} = 20/240 = .0833$

$$z = \frac{\hat{p} - p_0}{\sqrt{\dfrac{p_0(1 - p_0)}{n}}} = \frac{.0833 - .1}{\sqrt{\dfrac{.1(1 - .1)}{240}}} = \frac{-.0167}{.01936} = -0.86$$

Step 3: *p*-value ≈ .39. This is the combined probability (area) to the left of $z = -0.86$ and to the right of $z = 0.86$. Table A.1 gives $P(z < -0.86) = .1949$. Because this is a two-sided test, *p*-value $= 2 \times .1949 = .3898$.
Note: An exact *p*-value based on the binomial distribution is given by Minitab as .395.
Step 4: Cannot reject the null hypothesis. The result is not statistically significant because the *p*-value is not smaller than .05, the usual standard for significance.
Step 5: We cannot conclude that the proportion of UCD students who are left-handed differs from the national proportion.

Minitab output for this exercise is shown below.

Output for Exercise 12.91						
Test of p = 0.1 vs p not = 0.1						
Sample	X	N	Sample p	95.0% CI	Z-Value	P-Value
1	20	240	0.083333	(0.048366, 0.118300)	-0.86	0.389

12.93 The answer will vary. An example is the hypothesis that equal proportions of mean and women are unemployed. The populations are men and women and the proportion of interest is the proportion unemployed.

12.95 **a.** H_0: $p = \dfrac{31}{365} = .0849$

H_a: $p > .0849$ (31/365)

p is the proportion of the population born in October (nine months after the cold January Max observed).

b. Step 2: It seems reasonable to assume the sample of birthdays is representative of the larger population of birthdays and the sample size is large enough so that both np_0 and $n(1-p_0)$ are greater than 10. Here, $n = 170$ and $p_0 = 31/365 = .0849$

The test statistic is $z = \dfrac{\text{Sample estimate - Null value}}{\text{Null standard error}} = \dfrac{\hat{p} - p_0}{\sqrt{\dfrac{p_0(1-p_0)}{n}}}$.

Sample estimate $= \hat{p} = \dfrac{22}{170} = .1294$ (proportion of sample born in October)

$z = \dfrac{\hat{p} - p_0}{\sqrt{\dfrac{p_0(1-p_0)}{n}}} = \dfrac{.1294 - .0849}{\sqrt{\dfrac{.0849(1-.0849)}{170}}} = \dfrac{.0445}{.02138} = 2.08$

Step 3: p-value = .0188. This is the probability (area) to the right of 2.08. By symmetry this equals the probability to the left of –2.08. Table A.1 gives P(z < –2.08) = .0188.

Note: An exact p-value based on the binomial distribution is given by Minitab as .031.

Step 4: Reject the null hypothesis. The result is statistically significant because the p-value is less than .05.

Step 5: Conclude that people are more likely to be born in October than they would be if all 365 days were equally likely.

Minitab can be used to find the z-statistic and corresponding p-value. See the Minitab Tip on page 475 of the text for guidance. The output for this exercise is

Output for Exercise 12.95b						
Test of p = 0.0849 vs p > 0.0849						
Sample	X	N	Sample p	95.0% Lower Bound	Z-Value	P-Value
1	22	170	0.129412	0.087067	2.08	0.019

12.97 **a.** $p =$ proportion of all people who suffer from this chronic pain that would experience temporary relief if taking the new medication.

H_0: $p = .7$ (new has same success rate as standard)

H_a: $p > .7$ (new has better success rate than standard)

The null hypothesis could also be written as H_0: $p \leq .7$.

b. $p =$ proportion of all students at the university who would pay for the textbook ordering service.

H_0: $p \leq .05$ (5% or fewer would pay for the service)

H_a: $p > .05$ (at least 5% would pay for the service)

The null hypothesis might also be written as H_0: $p = .05$.

117

12.99 For $z > 0$, the p-value will be greater than .5. Because the alternative hypothesis is $p < p_0$, the p-value is the probability (area) *to the left* of the z-statistic. (See the first row of Table 12.1, page 465.) This area is greater than .5 when the value of z is positive.

12.101 Step 1: H_0: $p = .1$ (proportion for random selection)
H_a: $p > .1$ (proportion higher than with random selection)
p = proportion of population who would pick the number 7
Step 2: It seems reasonable to assume these students represent a larger population of students for the question of interest. The sample size is large enough so that both np_0 and $n(1 - p_0)$ are greater than 10. Here, $n = 190$ and $p_0 = .1$.

The test statistic is $z = \dfrac{\text{Sample estimate - Null value}}{\text{Null standard error}} = \dfrac{\hat{p} - p_0}{\sqrt{\dfrac{p_0(1 - p_0)}{n}}}$

Sample estimate = $\hat{p} = 56/190 = .2947$

$$z = \frac{\hat{p} - p_0}{\sqrt{\dfrac{p_0(1 - p_0)}{n}}} = \frac{.2947 - .10}{\sqrt{\dfrac{.1(1 - .1)}{190}}} = \frac{.1947}{.027164} = 8.95$$

Step 3: p-value ≈ 0. The p-value here is determined as $P(z > 8.95)$. $z = 8.95$ is "off the chart" for any method of finding this probability. There is virtually no probability that z could be larger than 8.95.
Step 4: We can reject the null hypothesis. The result is statistically significant.
Step 5: We can conclude that the proportion of the population who would pick the number 7 is greater than .1.

Minitab can be used to find the z-statistic and corresponding p-value. See the Minitab Tip on page 475 of the text for guidance. The output for this exercise is

Output for Exercise 12.101					
Test of p = 0.1 vs p > 0.1					
Sample	X	N	Sample p	95.0% Lower Bound	Z-Value	P-Value
1	56	190	0.294737	0.240331	8.95	0.000

12.103 Step 1: H_0: $p = .5$ (equal preference for two brands)
H_a; $p \neq .5$ (unequal preference for two brands)
p = proportion of population who prefer brand P
Step 2: In the problem statement, the given assumption is that these students represent the larger population of UC Davis students for the question of interest. The sample size is large enough so that both np_0 and $n(1 - p_0)$ are greater than 10. Here, $n = 159$ and $p_0 = .5$.

The test statistic is $z = \dfrac{\text{Sample estimate - Null value}}{\text{Null standard error}} = \dfrac{\hat{p} - p_0}{\text{Null standard error}}$

$\hat{p} = 80/159 = .503$. Null standard error $= \sqrt{\dfrac{p_0(1 - p_0)}{n}} = \sqrt{\dfrac{.5(1 - .5)}{159}} = .03965$

$$z = \frac{\hat{p} - p_0}{\text{Null standard error}} = \frac{.503 - .500}{.03965} = \frac{.003}{.03965} = 0.08$$

Step 3: p-value $\approx .94$. This is the combined probability that z is greater than 0.08 and less than −0.08. Table A.1 can be used to find $P(z < −0.08) = .4681$. The p-value is $2 \times .4681 = .9362$.
Step 4: We cannot reject the null hypothesis. The result is not statistically significant.
Step 5: We cannot conclude that the proportion within the population that prefers either brand is different from .5. In other words, there is not evidence that one drink or the other is more preferred.

118

Minitab can be used to find the *z*-statistic and corresponding *p*-value. See the Minitab Tip on page 475 of the text for guidance. The output for this exercise is shown below.

Output for Exercise 12.103						
Test of p = 0.5 vs p not = 0.5						
Sample	X	N	Sample p	95.0% CI	Z-Value	P-Value
1	80	159	0.503145	(0.425429, 0.580861)	0.08	0.937

12.105 Step 1: H_0: $p = .5$ (no preference for first drink)
        H_a: $p > .5$ (first drink is preferred)
p = proportion of population who would prefer first drink
Step 2: It seems reasonable to assume these students represent the larger population of UC Davis students for the question of interest, as was done in the previous exercise. The sample size is large enough so that both np_0 and $n(1-p_0)$ are greater than 10. Here, $n =159$ and $p_0 = .5$.

The test statistic is $z = \dfrac{\text{Sample estimate - Null value}}{\text{Null standard error}} = \dfrac{\hat{p} - p_0}{\sqrt{\dfrac{p_0(1-p_0)}{n}}}$.

Sample estimate = $\hat{p} = 86/159 = .541$.

$$z = \frac{\hat{p} - p_0}{\sqrt{\dfrac{p_0(1-p_0)}{n}}} = \frac{.541 - .5}{\sqrt{\dfrac{.5(1-.5)}{159}}} = \frac{.041}{.03965} = 1.03$$

Step 3: *p*-value ≈ .1515. This is the probability that *z* is greater than 1.03. By symmetry, $P(z > 1.03) = P(z < -1.03) = .1515$. Table A.1 can be used to find this.
Note: An exact *p*-value based on the binomial distribution is given as .171 by Minitab.
Step 4: Cannot reject the null hypothesis. The result is not statistically significant.
Step 5: Based on these data, we cannot conclude that more than one-half of the population would prefer the first drink presented to them.
Minitab can be used to find the *z*-statistic and corresponding *p*-value. See the Minitab Tip on page 475 of the text for guidance. The output for this exercise is shown below.

Output for Exercise 12.105						
Test of p = 0.5 vs p > 0.5						
Sample	X	N	Sample p	95.0% Lower Bound	Z-Value	P-Value
1	86	159	0.540881	0.475876	1.03	0.151

12.107 **a.** H_0: $p_1 - p_2 = 0$, or equivalently, $p_1 = p_2$
     H_a: $p_1 - p_2 \neq 0$, or equivalently, $p_1 \neq p_2$
p_1 = proportion of adult Canadians in February 2000 who favor marriages between people of the same sex , and p_2 = proportion of adult Canadians in April 1999 who favor marriages between people of the same sex
b. Step 2: Sample sizes are sufficiently large so that observed counts in both categories (favored or not) are greater than 10 for both years. Assume that the samples represent random samples from the populations in the two years, and that the samples were independently selected.
Test statistic is given in the output as $z = 3.19$
Steps 3. 4, and 5: *p*-value = .001. Reject the null hypothesis. Conclude that the proportion of adult Canadians favoring marriages between people of the same sex was different in February 2000 than in April 1999.
Note: It is not stated in the problem, but the reason this was of interest to researchers was provided in the article referenced in the problem. In February 2000 "The Federal Liberals, the party currently in power in Canada, have recently proposed a bill to extend legal benefits and obligations of married couples to common law couples regardless of sexual orientation." The issue was whether the publicity generated by the bill had changed the proportion in favor, and the hypothesis test shows that it probably did so.

119

c. The confidence interval does not cover 0, so it is evidence in favor of the alternative hypothesis.

12.109 **a.** Step 1: H_0: $p_1 - p_2 = 0$, or equivalently, $p_1 = p_2$ (proportions are equal)

H_a: $p_1 - p_2 < 0$, or equivalently, $p_1 < p_2$ (proportion smaller if taking aspirin)

p_1 = proportion that would have heart attack in population of men if they regularly take aspirin.
p_2 = proportion that would have heart attack in population of men if they regularly take placebo.
Step 2: There are two independent samples, which we assume to be random samples that represent larger populations. The sample sizes are sufficiently large so that observed counts in both categories (heart attack or not) are greater than 10 for both groups (aspirin and placebo).

Test statistic is $z = \dfrac{\text{Sample statistic - Null value}}{\text{Null standard error}} = \dfrac{-.007706 - 0}{.001541} = -5.00$. Details are:

Aspirin, $\hat{p}_1 = \dfrac{104}{11,037} = .009423$; placebo, $\hat{p}_2 = \dfrac{189}{11,034} = .017129$

Sample statistic $= \hat{p}_1 - \hat{p}_2 = .00942 - .01713 = -.007706$

Combined $\hat{p} = \dfrac{104 + 189}{11,037 + 11,034} = \dfrac{293}{22,071} = .01328$

Null standard error $= \sqrt{\dfrac{\hat{p}(1-\hat{p})}{n_1} + \dfrac{\hat{p}(1-\hat{p})}{n_2}} = \sqrt{\dfrac{.01328(1-.01328)}{11037} + \dfrac{.01328(1-.01328)}{11034}} = .001541$

Step 3: p-value $< .000001$ (or ≈ 0). It is the area to the left of $z = -5.00$ under a standard normal curve. At the bottom of the left page in Table A.1, the cumulative probability given for $z = -4.75$ is .000001. The one-sided p-value must be less than that probability because -5.00 is more extreme than -4.75.
Steps 4 and 5: We can reject the null hypothesis. The conclusion is that in the population(s) represented by the sample(s) the proportion that would have a heart attack is less if men were to take aspirin daily than it would be if men were to take a placebo daily.
b. The consequence of a type 1 error would be that people would take aspirin although it doesn't help. The consequence of a type 2 error would be that people would not take aspirin and heart attacks that could be prevented would not be. Assuming that taking aspirin doesn't have serious side effects, a type 2 error is more serious in this situation.

12.111 Step 1: H_0: $p \leq .50$ (not better than chance level)

H_a: $p > .50$ (predict better than chance level)

p = proportion of population of pregnant women that can predict the sex of their babies
Step 2: We must assume the sample represents a random sample from the population of pregnant women. The sample size is sufficiently large so that $n\hat{p}$ and $n(1-\hat{p})$ are both greater than 10.

Sample proportion correct guesses is $\hat{p} = \dfrac{57}{104} = .548$

Test statistic is $z = \dfrac{\text{Sample statistic - Null value}}{\text{Null standard error}} = \dfrac{\hat{p} - p_0}{\sqrt{\dfrac{p_0(1-p_0)}{n}}} = \dfrac{.548 - .50}{\sqrt{\dfrac{.50(1-.50)}{104}}} = \dfrac{.048}{.049} = 0.98$.

Step 3: p-value $= .1635$. It is the area (probability) to the right of $z = 0.98$.
$P(z > 0.98) = P(z \leq -0.98) = .1635$. Equivalently, $P(z > 0.98) = 1 - P(z \leq 0.98) = 1 - .8365 = .1635$.
Steps 4 and 5: We do not reject the null hypothesis for $\alpha = .05$. There is not enough evidence to conclude that in the population of pregnant women represented by the sample, the proportion able to predict the sex of their babies is higher than .50 (the chance level).
Note: Minitab could be used to do this exercise (see Minitab Tip on page 475). The program reports the exact I-value based on binomial distribution probabilities as .189.

12.113 Step 1: H_0: $p \leq .50$ (not a majority)

H_a: $p > .50$ (a majority were dissatisfied)

120

p = proportion of U.S. adults dissatisfied with K-12 education in August 2000
Step 2: The sample was randomly selected from the population of U.S. adults and the sample size is sufficiently large so that np_0 and $n(1-p_0)$ are both greater than 10.

Sample proportion dissatisfied is $\hat{p} = \dfrac{622}{1019} = .6104$

Test statistic is $z = \dfrac{\text{Sample statistic - Null value}}{\text{Null standard error}} = \dfrac{\hat{p} - p_0}{\sqrt{\dfrac{p_0(1-p_0)}{n}}} = \dfrac{.6104 - .50}{\sqrt{\dfrac{.50(1-.50)}{1019}}} = \dfrac{.1104}{.01566} = 7.05$.

Step 3: p-value ≈ 0. It is the area (probability) to the right of $z = 7.05$. This z-value is beyond the last value given in the "In the Extreme" section of Table A.1, so the area to the right must be nearly 0.
Steps 4 and 5: We can reject the null hypothesis. The conclusion is that in the population of U.S. adults in August 2000, a majority were dissatisfied with the quality of K-12 education.
Note: Minitab could be used to do this exercise (see Minitab Tip on page 475). The exact binomial p-value is given as .000.

12.115 **a.** Step 1: H_0: $p \le .50$ (not a majority)
$\quad\quad\quad\quad\quad$ H_a: $p > .50$ (a majority were dissatisfied)
p = proportion of U.S. adults dissatisfied with K-12 education in August 1999
Step 2: The sample was randomly selected from the population of U.S. adults and the sample size is sufficiently large so that np_0 and $n(1-p_0)$ are both greater than 10 (and in fact are both the same at 1028/2 = 514).

Sample proportion dissatisfied is $\hat{p} = \dfrac{524}{1028} = .5097$

Test statistic is $z = \dfrac{\text{Sample statistic - Null value}}{\text{Null standard error}} = \dfrac{\hat{p} - p_0}{\sqrt{\dfrac{p_0(1-p_0)}{n}}} = \dfrac{.5097 - .50}{\sqrt{\dfrac{.50(1-.50)}{1028}}} = \dfrac{.0097}{.0156} = 0.62$

Step 3: p-value $= .2676$. It is the area (probability) to the right of $z = 0.62$.
$P(z > 0.62) = P(z \le -0.62) = .2676$. Equivalently, $P(z > 0.62) = 1 - P(z \le 0.62) = 1 - .7324 = .2676$. Minitab gives the p-value as .266. This minor difference occurs because Minitab keeps track of more decimal places in the z-statistic.
Steps 4 and 5: We do not reject H_0 for $\alpha = .05$. The conclusion concerning the population of U.S. adults in August 1999 is that we cannot say that a majority were dissatisfied with the quality of K-12 education.
Note: Minitab could be used to do this exercise (see Minitab Tip on page 475). The program reports the exact p-value based on binomial distribution probabilities as 0.277.

b. Step 1: H_0: $p_1 - p_2 = 0$, or equivalently, $p_1 = p_2$
$\quad\quad\quad\quad$ H_a: $p_1 - p_2 \ne 0$, or equivalently, $p_1 \ne p_2$

p_1 = proportion of population of U.S. adults dissatisfied with K-12 education in August 2000
p_2 = proportion of population of U.S. adults dissatisfied with K-12 education in August 1999
Step 2: We assume two independent samples that represent the populations in the two different years. The observed counts in both categories (dissatisfied or otherwise) are greater than 10 for both years.

Test statistic is $z = \dfrac{\text{Sample statistic - Null value}}{\text{Null standard error}} = \dfrac{.1007 - 0}{.02192} = 4.59$. Details are:

August 2000, $\hat{p}_1 = \dfrac{622}{1019} = .6104$; August 1999, $\hat{p}_2 = \dfrac{524}{1028} = .5097$

Sample statistic $= \hat{p}_1 - \hat{p}_2 = .6104 - .5097 = .1007$

Combined $\hat{p} = \dfrac{622 + 524}{1019 + 1028} = \dfrac{1146}{2047} = .5598$

121

Null standard error = $\sqrt{\dfrac{\hat{p}(1-\hat{p})}{n_1}+\dfrac{\hat{p}(1-\hat{p})}{n_2}} = \sqrt{\dfrac{.5598(1-.5598)}{1019}+\dfrac{.5598(1-.5598)}{1028}} = .02192$

Step 3: p-value ≈ 0. It is the combined area (probability) to the right of $z = 4.59$ and to the left of -4.59. With appropriate software or calculator, it can be determined as $2 \times P(z < -4.59) =$
$2 \times .00002 = .00004$. Using Table A.1, the p-value would be estimated as p-value $< (2 \times .00001)$, based on the cumulative probability given in Table A.1 for $z = -4.26$.

Steps 4 and 5: We can reject the null hypothesis. The conclusion is that the proportion of U.S. adults dissatisfied with the quality of K-12 education changed from August 1999 to August 2000.

12.117 Step 1: H_0: $p_1 - p_2 = 0$, or equivalently, $p_1 = p_2$
  H_a: $p_1 - p_2 \neq 0$, or equivalently, $p_1 \neq p_2$

p_1 = proportion favoring legalization of marijuana in the U.S. population of men in 2008
p_2 = proportion favoring legalization of marijuana in the U.S. population of women in 2008

Step 2: The sample represents a random sample from the U.S. population and the sample size is sufficiently large so that observed counts in both categories (legal or not legal) are greater than 10 in both groups (males and females). Minitab output is given below.

```
                    Output for Exercise 12.117
sex       X    N   Sample p
Female   255  691  0.369030
Male     241  556  0.433453

Difference = p (Female) - p (Male)
Estimate for difference:  -0.0644228
95% CI for difference:  (-0.119114, -0.00973155)
Test for difference = 0 (vs not = 0):  Z = -2.31   P-Value = 0.021
```

Step 2 continued and Steps 3, 4, and 5: The test statistic is $z = -2.31$ and the p-value is .021. Reject the null hypothesis and conclude that in the U.S. population in 2008, different proportions of males and females favored legalization of marijuana. Note that the observed proportion favoring legalization is higher for males ($\hat{p}_1 \approx .43$) than for females ($\hat{p}_2 \approx .37$).

Note: One strategy for using Minitab is to first do a two-way table to find the relevant summary counts and to then enter these counts as "Summarized Data" in Minitab's test for 2 proportions (see the Minitab tip on page 480). This will allow you to choose the category "legal" as the category of interest, rather than letting Minitab choose which category to use.

CHAPTER 13
ODD-NUMBERED SOLUTIONS

13.1　**a.** Not true. The correct conclusion if the null hypothesis cannot be rejected is to "fail to reject the null hypothesis."
b. True.

13.3　**a.** The difference between the means of two populations.
b. The population mean of paired differences.

13.5　**a.** The parameter of interest is $\mu_1 - \mu_2$ where μ_1 = the mean height of the population of 25-year-old women and μ_2 = the mean height of the population of 45-year-old women.
b. The parameter of interest is μ_d = the mean difference in number of minutes the two flights of interest are late, for the population of all days over several years.

13.7　Refer to the solutions to 13.5 for the definitions of the parameters.
a. $H_0: \mu_1 - \mu_2 = 0$
b. $H_0: \mu_d = 0$

13.9　A standardized statistic is used because we need some way to know if the difference between the sample statistic and the null value is larger than what would be expected by chance. Just knowing the difference without any reference to how large it would be by chance doesn't tell us much.

13.11　**a.** The term "hypothesis testing" is used because we state two hypotheses about the truth in the population and use data to assess them.
b. The term "significance testing" is used because we assess whether an observed difference between sample data and our null hypothesis value is substantial enough, or significant enough, to lead us to reject the null hypothesis value as the truth.

13.13　Yes, The hypotheses can (and should) be specified before collecting the data.

13.15　The parameter of interest is μ = mean percentage of net patient revenue spent on charity care for the population of all nonprofit hospitals. The hypotheses are $H_0: \mu \geq 4\%$ and $H_a: \mu < 4\%$.

13.17　$t = \dfrac{(270 - 250)}{5} = 4.00$ (Note that the standard error, not the standard deviation, is given as 5.)

13.19　**a.** The *p*-value is .048. df $= n - 1 = 81 - 1 = 80$. In Table A.3, under $t = 2.0$, the one-sided *p*-value is given as .024. The two-sided p-value is 2(.024) = .048.

Figure for Exercise 13.19a

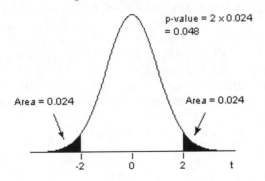

b. The p-value = .048, as it was in part (a), and the figure will be the same as for part (a).

13.21 **a.** Rejection region is $t \geq 1.83$; reject H_0. Use one-tailed $\alpha = .05$ column and df = 9 row.
b. Rejection region is $t \leq -1.83$; reject H_0. Use one-tailed $\alpha = .05$ column and df = 9 row.
c. Rejection region is $t \leq -1.83$; do not reject H_0. Use one-tailed $\alpha = .05$ column and df = 9 row.
d. Rejection region is $|t| \geq 2.26$; do not reject H_0. Use two-tailed $\alpha = .05$ column and df = 9 row.

13.23 **a.** Null standard error is $\text{s.e.}(\bar{x}) = \dfrac{s}{\sqrt{n}} = \dfrac{15}{\sqrt{50}} = 2.121$

b. $\dfrac{\text{Sample statistic - Null value}}{\text{Null standard error}} = \dfrac{102 - 100}{2.121} = 0.943$

c. No, to find the proper p-value we must know whether the alternative hypothesis is one-sided or two-sided.

13.25 **a.** No. The mean is 8 minutes and the standard deviation is 10 minutes. Bell-shaped data would range from about three standard deviations below the mean to three standard deviations above the mean. In this case, even one standard deviation below the mean is negative, and it isn't possible to have calls of negative length.

b. Yes. The sample size is large (200) so it doesn't matter that the data are not bell-shaped.

c. Step 1: $H_0: \mu = 9.2$ versus $H_a: \mu \neq 9.2$, where μ = mean length of calls for all seniors.
Step 2: Sample size (200) is sufficiently large to proceed. It is stated in the problem that we can assume the sample represents a random sample of the population of all seniors.

Test statistic is $t = \dfrac{\text{Sample statistic - Null value}}{\text{Null standard error}} = \dfrac{\bar{x} - \mu_0}{\dfrac{s}{\sqrt{n}}} = \dfrac{8 - 9.2}{\dfrac{10}{\sqrt{200}}} = \dfrac{-1.2}{0.7071} = -1.70$

Steps 3, 4, and 5: df = $n - 1 = 200 - 1 = 199$. The p-value is $2 \times P(t < -1.70)$. Using Table A.3, the p-value is between $2(.037) = .074$ and $2(.051) = .102$. With software like Excel or Minitab, or a suitable calculator, it can be found that the p-value is about $2(.045) = .09$. Using the standard $\alpha = .05$, we cannot reject the null hypothesis. There is not sufficient evidence to conclude that the mean length of calls for seniors is different from the mean for the general population.

13.27 Step 1: $H_0: \mu = 9.2$ versus $H_a: \mu \neq 9.2$, where μ = mean length of calls for all seniors.
Step 2: Sample size (200) is sufficiently large to proceed. It is stated in the problem that we can assume the sample represents a random sample of the population of all seniors.

Test statistic is $t = \dfrac{\text{Sample statistic - Null value}}{\text{Null standard error}} = \dfrac{\bar{x} - \mu_0}{\dfrac{s}{\sqrt{n}}} = \dfrac{8 - 9.2}{\dfrac{10}{\sqrt{200}}} = \dfrac{-1.2}{0.7071} = -1.70$

Steps 3, 4, and 5: df = 199. Rejection region is $|t| \geq 1.98$, found in the 100 row of Table A.2, reading up from the column at the bottom for *Two-tailed* α of .05. (Software gives the value for 199 df as 1.971.) We cannot reject the null hypothesis because the test statistic $t = -1.70$ is not in the rejection region. There is not sufficient evidence to conclude that the mean length of calls for seniors is different from the mean for the general population.

13.29 Step 1: $H_0: \mu = 4.7$, $H_a: \mu < 4.7$, where μ = mean time to graduate (in years) for students who participated in the honors program in their first year of college.
Step 2: Sample size (30) is sufficiently large to proceed. It is stated in the problem that we can assume the sample represents a random sample of the population of students who participated in the honors program in their first year of college.

Test statistic is $t = \dfrac{\text{Sample statistic - Null value}}{\text{Null standard error}} = \dfrac{\bar{x} - \mu_0}{\dfrac{s}{\sqrt{n}}} = \dfrac{4.5 - 4.7}{\dfrac{0.5}{\sqrt{30}}} = \dfrac{-0.2}{0.0913} = -2.19$

Steps 3, 4, and 5: df = $n - 1 = 30 - 1 = 29$. The p-value is $P(t < -2.19)$. Using Table A.3, the p-value is between .013 and .027. With software like Excel or Minitab, or a suitable calculator, it can be found that

124

the *p*-value is .018. Using the standard $\alpha = .05$, we can reject the null hypothesis and conclude that the mean time to graduate is lower for the population of students who participated in the honors program than it is for the general population of students.

13.31 <u>Step 1</u>: $H_0:\mu = 65$, $H_a:\mu < 65$, where μ = mean height (in inches) for the population of college women who prefer to sit in the front of the class.

<u>Step 2</u>: We will assume that the women in the sample are representative of all women in the population when it comes to height. The sample size (38) is large enough to continue, but a graph to check for skewness or outliers is a good idea for a moderate sample size. The histogram below shows that there are no extreme outliers or skewness.

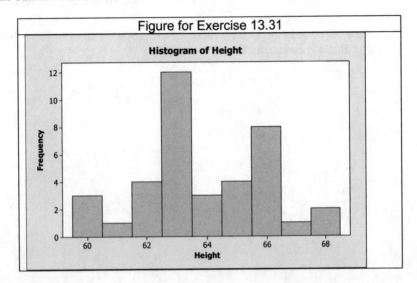

The summary statistics are $\bar{x}$ = 63.855 inches and s = 2.086 inches. The test statistic is

$$t = \frac{\text{Sample statistic - Null value}}{\text{Null standard error}} = \frac{\bar{x} - \mu_0}{\frac{s}{\sqrt{n}}} = \frac{63.855 - 65}{\frac{2.086}{\sqrt{38}}} = \frac{-1.145}{0.3383} = -3.38$$

<u>Steps 3, 4, and 5</u>: df $= n - 1 = 38 - 1 = 37$. The *p*-value is $P(t < -3.38)$. Using Table A.3, all we can say is that the *p*-value is less than .003, using the entry for df = 30 and the column for $t = 3.00$. Using Minitab the *p*-value is given as .001. Using the standard $\alpha = .05$, we can reject the null hypothesis and conclude that the mean height for women who prefer to sit in the front of the class is lower than it is for the population in general. In other words, women who prefer to sit in the front of the class are shorter on average than the general population.

13.33 The two tests are conceptually different because the paired *t*-test is concerned with comparing two means whereas a one-sample *t*-test is concerned with comparing one mean to a fixed value. However, once the sample differences have been computed for paired data, the two tests are computationally exactly the same.

13.35 $t = \dfrac{(-4 - 0)}{15/\sqrt{50}} = -1.89$, df = 49, *p*-value is between $2 \times .04 = .08$ and $2 \times .026 = .052$, using Table A.3 with df = 40. Using df = 50, $.05 < p\text{-value} < .078$. Exact *p*-value (using software) is $2 \times .0323 = .0646$.

13.37 a. $H_0: \mu_d = 0$

$H_a: \mu_d > 0$ (on average, student height > mom's height)

μ_d = mean "student height–mom's height" difference for population of college student represented by the sample

Note: The null hypothesis could also be written as $H_0: \mu_d \leq 0$

125

b. $t = \dfrac{\text{Sample statistic - Null value}}{\text{Null standard error}} = \dfrac{1.285 - 0}{0.2745} = 4.68$

Sample statistic is observed mean difference in heights, $\bar{d} = 1.285$ inches.

Null value is $\mu_d = 0$

Null standard error $= \dfrac{s_d}{\sqrt{n}} = \dfrac{2.647}{\sqrt{93}} = 0.2745$ (given in output)

c. df $= n - 1 = 93 - 1 = 92$

d. Reported p-value $= 0.000$. Reject the null hypothesis. For the population of college women represented by the sample, we can conclude that, on average, students' heights are greater than their mother's heights. The observed magnitude of the difference is $\bar{d} = 1.282$ inches.

e. The p-value is the area (probability) to the right of 4.68 in a t-distribution with df $= 92$. It is obvious from the figure below that it is essentially 0.

Figure for Exercise 13.37e

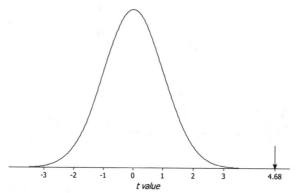

13.39   Step 1: H$_0$: $\mu_d = 0$

H$_a$: $\mu_d > 0$ (January weight is greater, on average)

μ_d = mean "January weight–November weight" difference for population represented by the sample.

Step 2: The sample size is sufficiently large to proceed. We must assume that the sample represents a random sample from a larger population.

Test statistic is $t = \dfrac{\text{Sample statistic - Null value}}{\text{Null standard error}} = \dfrac{0.37 - 0}{0.109} = 3.39$

Null standard error $= \dfrac{s_d}{\sqrt{n}} = \dfrac{1.52}{\sqrt{195}} = 0.109$ kg.

Step 3: p-value ≈ 0. It is the area (probability) to the right of $t = 3.39$ in a t-distribution with df $= n - 1 = 195 - 1 = 194$. With Table A.3, it can be estimated from the df $= 100$ row that the p-value is less than .002. With appropriate software of calculator, it can be found that $P(t > 3.39) = .0004$.

Steps 4 and 5: Reject the null hypothesis. The conclusion about the population is that the mean difference between January and November weights is greater than 0. In the sample, the observed magnitude of the difference was $\bar{d} = 0.37$ kg ≈ 0.8 pounds. (Note that this average gain of less than one pound may not have much *practical* importance.)

13.41   Step 1: H$_0$: $\mu_d = 0$

H$_a$: $\mu_d > 0$ (January weight is greater, on average)

μ_d = mean "January weight–November weight" difference for population represented by the sample.

Step 2: The sample size is sufficiently large to proceed. We must assume that the sample represents a random sample from a larger population.

126

Test statistic is $t = \dfrac{\text{Sample statistic - Null value}}{\text{Null standard error}} = \dfrac{0.37 - 0}{0.109} = 3.39$

Null standard error $= \dfrac{s_d}{\sqrt{n}} = \dfrac{1.52}{\sqrt{195}} = 0.109 \, \text{kg}$.

Step 3: df $= 195 - 1 = 194$. The test is one-sided and $\alpha = .01$. In Table A.2 use the row for df $= 100$ (the closest to 194 in the Table) and the column for one-tailed $\alpha = .01$ to find the critical value $t^* = 2.36$. So the rejection region is $t \geq 2.36$. Using software and df $= 194$ it is $t \geq 2.35$.

Steps 4 and 5: Reject the null hypothesis because the test statistic $t = 3.39$ is in the rejection region . The conclusion about the population is that the mean difference between January and November weights is greater than 0. In the sample, the observed magnitude of the difference was $\bar{d} = 0.37 \, \text{kg} \approx 0.8$ pounds. (Note that this average gain of less than one pound may not have much *practical* importance.)

13.43 **a.** The parameter of interest is μ_d = mean weight loss (in pounds) in the first three weeks of following a diet plan for the population of all people who would follow the plan.

b. $H_0: \mu_d = 10$, $H_a: \mu_d < 10$

c. Test statistic is $t = \dfrac{\text{Sample statistic - Null value}}{\text{Null standard error}} = \dfrac{8 - 10}{\dfrac{4}{\sqrt{20}}} = -2.236$.

d. df $= 20 - 1 = 19$; p-value is the area (probability) to the left of -2.236 in a t-distribution with df $= 19$. From Table A.3 we find that the p-value is between .015 and .030. Software or an appropriate calculator can be used to determine that the p-value is .0188.

e. The consumer advocacy group can reject the null hypothesis and conclude that the mean weight loss for the first three weeks of this diet plan is less than 10 pounds.

13.45 **a.** The parameter of interest is μ_d = mean difference in blood pressure while waiting to see a dentist and an hour after visiting the dentist (waiting bp – later bp).

b. $H_0: \mu_d = 0$

$H_a: \mu_d > 0$ (Blood pressure higher on average while waiting for dentist than an hour after.)

c. First, find the differences (Before – After). The 10 differences are: 14, –2, 9, –6, 12, 5, 6, 8, 4, 7.

The mean of the differences is $\bar{d} = 5.7$ and the standard deviation is $s_d = 6.0194$.

A stem-and-leaf plot and the details of the test are shown in the Minitab output below.

Minitab output and plot for Exercise 13.45c					
Paired T for Waiting – After				Stem-and-leaf of	
				Difference N = 10	
	N	Mean	StDev	SE Mean	Leaf Unit = 1.0
Waiting	10	127.800	10.963	3.467	
After	10	122.100	13.034	4.122	1 -0 6
Difference	10	5.70000	6.01941	1.90351	2 -0 2
					3 0 4
95% lower bound for mean difference: 2.21066					(5) 0 56789
T-Test of mean difference = 0 (vs > 0): T-					2 1 24
Value = 2.99 P-Value = 0.008					

Step 2: Assume that the sample represents the larger population of people who visit a dentist. The stem-and-leaf plot shows that there are no outliers or skewness, so a t-test is appropriate. The test statistic is shown in the output as $t = 2.99$. Computed by hand it is

$t = \dfrac{\text{Sample statistic-Null value}}{\text{Null standard error}} = \dfrac{5.7 - 0}{\dfrac{6.0194}{\sqrt{10}}} = \dfrac{5.7}{1.9035} = 2.99$.

Step 3: The p-value is shown in the output as .008. From software it is .0076 (using df $= 9$) and from Table A.3 it is between .015 and .007.

<u>Steps 4 and 5</u>: Reject the null hypothesis because p-value < .05. Conclude that on average, blood pressure is higher while waiting to see a dentist than it is an hour after the visit.

13.47   **a.** H_a: $\mu_1 - \mu_2 < 0$.
      b. H_a: $\mu_1 - \mu_2 < 0$.

13.49   **a.** Paired.
      b. Two-sample.

13.51   **a.** $t = \dfrac{(35-33)-0}{\sqrt{\dfrac{10^2}{100}+\dfrac{9^2}{81}}} = \dfrac{2}{\sqrt{2}} = 1.414$
      b. $t = 48/22 = 2.18$

13.53   **a.** H_0: $\mu_1 - \mu_2 = 0$, H_a: $\mu_1 - \mu_2 \neq 0$
      b. $t = 2.78$, reported in the output as "T-value."
      c. $t = \dfrac{(98.102-97.709)-0}{\sqrt{\dfrac{(0.651)^2}{46}+\dfrac{(0.762)^2}{54}}} = \dfrac{0.393}{0.1413} = 2.78$

      d. Reject the null hypothesis because the p-value of .007 is less than .05. Conclude that men and women do not have equal mean body temperatures.

13.55   <u>Step 1</u>: H_0: $\mu_1 - \mu_2 = 0$, or equivalently $\mu_1 = \mu_2$
               H_a: $\mu_1 - \mu_2 \neq 0$, or equivalently, $\mu_1 \neq \mu_2$

       μ_1 = mean weight for population of Cambridge rowers

       μ_2 = mean weight for population of Oxford rowers

       <u>Step 2</u>: A dotplot (as well as a boxplot) shows that there are no outliers and the data are reasonably symmetric.

Figure for Exercise 13.55

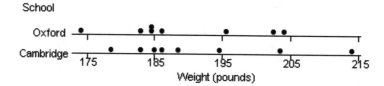

For the unpooled procedure, $t = 0.43$ (df ≈ 13, reported by Minitab).

Output for Exercise 13.55

```
Two-sample T for Cambridge vs Oxford

          N      Mean     StDev    SE Mean
cambridg  8      191.6    11.8       4.2
Oxford    8      189.3    10.4       3.7


Difference = mu Cambridge - mu Oxford
Estimate for difference:  2.38
95% CI for difference: (-9.67, 14.42)
T-Test of difference = 0 (vs not =): T-Value = 0.43   P-Value =
0.677  DF = 13
```

Steps 3, 4, and 5: *p*-value = .677 (from output). We do not reject the null hypothesis for any reasonable value of α. There is not sufficient evidence to conclude that the populations of Cambridge and Oxford rowers differ with regard to mean weight.

Note: Because the sample sizes are equal the value of the test statistic for the pooled procedure is identical; see the first bullet on page 518. The degrees of freedom would be slightly smaller but the conclusion would be the same. Either method is appropriate in this case.

13.57 **a.** Step 1: $H_0: \mu_1 - \mu_2 = 0$, or equivalently $\mu_1 = \mu_2$

$H_a: \mu_1 - \mu_2 \neq 0$, or equivalently, $\mu_1 \neq \mu_2$

μ_1 = mean weight loss for population of sedentary men if they were to diet

μ_2 = mean weight loss for population of sedentary men if they were to exercise

Step 2: The sample sizes are sufficiently large to proceed. We must assume the samples represent random samples from larger populations of men who would either diet or exercise to lose weight.

For unpooled procedure,

Test statistic is $t = \dfrac{\text{Sample statistic - Null value}}{\text{Null standard error}} = \dfrac{3.2 - 0}{0.806} = 3.97$

Sample statistic is $\bar{x}_1 - \bar{x}_2 = 7.2 - 4.0 = 3.2$ kg.

Standard error is $\sqrt{\dfrac{s_1^2}{n_1} + \dfrac{s_2^2}{n_2}} = \sqrt{\dfrac{3.7^2}{42} + \dfrac{3.9^2}{47}} = \sqrt{0.6496} = 0.806$

For pooled procedure, $t = \dfrac{3.2 - 0}{0.808} = 3.96$.

$s_p = \sqrt{\dfrac{(42-1)3.7^2 + (47-1)3.9^2}{42+47-2}} = \sqrt{14.49} = 3.81$, and pooled standard error is

$s.e.(\bar{x}_1 - \bar{x}_2) = s_p \sqrt{\dfrac{1}{n_1} + \dfrac{1}{n_2}} = 3.81 \sqrt{\dfrac{1}{42} + \dfrac{1}{47}} = 0.808$

Steps 3, 4, and 5: *p*-value ≈ .0002 for either procedure. It is calculated as $2 \times P(t > 3.97)$ for the unpooled procedure, and as $2 \times P(t > 3.96)$ for the pooled. For unpooled, df ≈ 41 (minimum of $n_1 - 1$ and $n_2 - 1$) and for pooled, df = $n_1 + n_2 - 2 = 42 + 47 - 2 = 85$. With Table A.3, the two-sided *p*-value would be estimated to be less than 2(.002) = .004 for both procedures.

Steps 4 and 5: Reject the null hypothesis. We can conclude that the mean weight loss would be different for the population of sedentary men if they were to diet compared to if they were to exercise.

b. In this situation we can use the pooled procedure because the two groups have similar sample sizes and standard deviations. But either procedure would be acceptable and the results are very similar.

13.59 Step 1: $H_0: \mu_1 - \mu_2 = 0$, or equivalently $\mu_1 = \mu_2$

$H_a: \mu_1 - \mu_2 \neq 0$, or equivalently, $\mu_1 \neq \mu_2$

μ_1 = mean sleep hours for population of UC Davis students

μ_2 = mean sleep hours for population of Penn State students

Step 2: The sample sizes are sufficiently large to proceed. We must assume the samples represent random samples from the larger populations of students at these schools.

For unpooled procedure,

Test statistic is $t = \dfrac{\text{Sample statistic - Null value}}{\text{Null standard error}} = \dfrac{-0.18 - 0}{0.192} = -0.94$

Sample statistic is $\bar{x}_1 - \bar{x}_2 = 6.93 - 7.11 = -0.18$ hours.

Standard error $\sqrt{\dfrac{s_1^2}{n_1} + \dfrac{s_2^2}{n_2}} = \sqrt{\dfrac{1.71^2}{173} + \dfrac{1.95^2}{190}} = \sqrt{0.0369} = 0.192$

129

For pooled procedure, $t = \dfrac{-0.18 - 0}{0.193} = -0.93$

$$s_p = \sqrt{\frac{(173-1)1.71^2 + (190-1)1.95^2}{173 + 190 - 2}} = \sqrt{3.365} = 1.834 \text{, and pooled standard error is}$$

$$s.e.(\bar{x}_1 - \bar{x}_2) = s_p \sqrt{\frac{1}{n_1} + \frac{1}{n_2}} = 1.834 \sqrt{\frac{1}{173} + \frac{1}{190}} = 0.193$$

Step 3: p-value $\approx .35$ for either procedure. It is calculated as $2 \times P(t > -0.94)$ for the unpooled procedure, and as $2 \times P(t > -0.93)$ for the pooled procedure. For unpooled, df ≈ 172 (minimum of $n_1 - 1$ and $n_2 - 1$) and for pooled, df $= n_1 + n_2 - 2 = 173 + 190 - 2 = 361$. With Table A.3, the two-sided p-value would be estimated to be greater than $2(.102) = .204$ for both procedures.

Steps 4 and 5: Do not reject the null hypothesis using any reasonable α. We do not reject the possibility that the mean hours of sleep are the same for the populations of students at the two schools.

13.61　**a.**　Step 1: $H_0\!: \mu_1 - \mu_2 = 0$, or equivalently $\mu_1 = \mu_2$

$H_a\!: \mu_1 - \mu_2 > 0$, or equivalently, $\mu_1 > \mu_2$ (mean speed higher with no seatbelt use)

$\mu_1 =$ mean speed under road conditions similar to those in the study for the population of drivers who don't wear seatbelts

$\mu_2 =$ mean speed under road conditions similar to those in the study for the population of drivers who do wear seatbelts

Step 2: We must assume the samples represent random samples from the larger populations of drivers. The sample sizes are small so plots should be used to determine if outliers or skewness are present. The histograms below show the speeds for the drivers without and with seatbelts, and do not exhibit any problems with outliers or skewness.

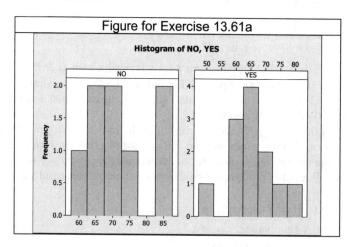

Figure for Exercise 13.61a

Using the unpooled procedure,

Test statistic is $t = \dfrac{\text{Sample statistic - Null value}}{\text{Null standard error}} = \dfrac{7.167 - 0}{3.795} = 1.89$. Details are:

Sample statistic is $\bar{x}_1 - \bar{x}_2 = 72.5 - 65.33 = 7.167$mph.

Standard error $\sqrt{\dfrac{s_1^2}{n_1} + \dfrac{s_2^2}{n_2}} = \sqrt{\dfrac{8.82^2}{8} + \dfrac{7.49^2}{12}} = 3.795$

Step 3: From Minitab, df $= 13$ and p-value $= .041$. Using Table A.3 and the conservative df $=$ smaller of sample sizes $- 1$, df $= 8 - 1 = 7$, p-value is between $.047$ and $.053$. Notice that this range differs from the p-value given by Minitab because the conservative degrees of freedom have been used.

130

Steps 4 and 5: Using the p-value of .041, the null hypothesis can be rejected. It can be concluded that the mean speed for the population of drivers who do not wear seatbelts is higher than the mean speed for those who do wear them.

b. Step 1 and the "checking the conditions" part of Step 2 are the same as in part (a). For the pooled procedure, $t = \dfrac{7.167 - 0}{3.665} = 1.96$. Details are:

$$s_p = \sqrt{\frac{(8-1)8.82^2 + (12-1)7.49^2}{8+12-2}} = \sqrt{64.536} = 8.03 \text{, and pooled standard error is}$$

$$s.e.(\overline{x}_1 - \overline{x}_2) = s_p \sqrt{\frac{1}{n_1} + \frac{1}{n_2}} = 8.03 \sqrt{\frac{1}{8} + \frac{1}{12}} = 3.665 .$$

Step 3: df $= 8 + 12 - 2 = 18$. From Minitab, p-value $= .033$. Using Table A.3 with df $= 18$, the p-value range is .044 to .030.

Steps 4 and 5: Reject the null hypothesis. Conclude that the mean speed for the population who do not wear seatbelts is higher than it is for the population of people who do wear them.

c. The results of the two procedures are similar, with test statistics of 1.89 and 1.96. The sample standard deviations are not too far apart (8.82 and 7.49) so it is not unreasonable to use the pooled procedure. However, the larger sample size (12) produced the smaller standard deviation so it is probably safer to use the unpooled procedure.

d. df $= 13$, so using Table A.2 with one-tailed $\alpha = .05$ the critical value is $t^* = 1.77$ and the rejection region is $t \geq 1.77$. The test statistic is 1.89, which is in the rejection region, so the null hypothesis can be rejected. Using the conservative df $= 7$, the critical value is $t^* = 1.89$ and the rejection region is $t \geq 1.89$. The test statistic is 1.89 (coincidentally the same as the critical value) so the null hypothesis can be rejected. We can conclude that the mean speed for the population of drivers who do not wear seatbelts is higher than it is for the population of drivers who do wear seat belts.

13.63 **a.** No. The null value is covered by the 95% confidence interval.
b. No. The null value is covered by the 90% confidence interval. (See first bullet on page 522.)
c. Yes. The entire 90% confidence interval falls above the null value, so the alternative hypothesis that $\mu > 25$ can be accepted. (See the second bullet on page 522.)

13.65 **a.** If the null value is covered by a 95% confidence interval, the null hypothesis is not rejected and the test is not statistically significant at level .05.
If the null value is not covered by a 95% confidence interval, the null hypothesis is rejected and the test is statistically significant at level .05.
b. If the null value is covered by a 99% confidence interval, the null hypothesis is not rejected and the test is not statistically significant at level .01.
If the null value is not covered by a 99% confidence interval, the null hypothesis is rejected and the test is statistically significant at level .01.

13.67 **a.** Reject H_0 for level of significance $\alpha = .05$. The null value (100) is not within the 95% confidence interval.
b. Do not reject H_0 for level of significance $\alpha = .025$. The entire interval falls into the null hypothesis region ($H_0: p \geq .10$). The *one-sided* level of significance corresponding to 95% confidence is $.05/2 = .025$.
c. Reject H_0 for level of significance $\alpha = .025$. The entire interval falls into the alternative hypothesis region. The *one-sided* level of significance corresponding to 95% confidence is $.05/2 = .025$.
d. Do not reject H_0 for level of significance $\alpha = .05$. The interval covers the null value (0).

13.69 **a.** Can't tell. The entire interval could be greater than 10 or it could include 10.
b. The interval would not include the value 10. If it did, the null hypothesis would not have been rejected.
c. The interval would include the value 10. Otherwise, the null hypothesis would have been rejected.
d. The interval would not include the value 10. If it did, the null hypothesis would not have been rejected.

13.71 **a.** There are two separate questions of interest. For the first one, p = proportion of all adults who have a fear of going to the dentist. (The proportion can be converted to a percent.) For the second one, μ = mean number of visits made to a dentist in the past 10 years for the population of adults who fear going to the dentist.

b. Confidence intervals would be appropriate for both parameters. There are no obvious null values to do a hypothesis test.

13.73 **a.** $\mu_1 - \mu_2$ where μ_1 and μ_2 are the mean running times for the 50-yard dash for the populations of first grade boys and girls, respectively.

b. Both. A confidence interval would be more appropriate to estimate the magnitude of the mean difference in running times. A hypothesis test could be used to determine if the mean times are significantly different.

13.75 **a.** It would be appropriate to estimate $p_1 - p_2$, the difference in proportions of college men and women who would answer yes, and also to test whether $p_1 - p_2 = 0$. You may also want to estimate p_1 and p_2 separately.

b. Test H_0: $\mu_1 - \mu_2 = 0$ versus H_a: $\mu_1 - \mu_2 < 0$, where μ_1 and μ_2 are the mean IQs for the populations of children whose mothers smoke at least 10 cigarettes a day during pregnancy and for those whose mothers don't smoke during pregnancy. Note that we may also want to estimate the difference in means, but the question posed is whether the IQs are lower for children of smokers, indicating that the primary interest is in testing that hypothesis rather than in estimating the difference.

13.77 **a.** Effect size is $\dfrac{2.24}{\sqrt{100}} = 0.224$; small.

b. Effect size is $\dfrac{-2.83}{\sqrt{50}} = -0.4$; closer to medium.

13.79 **a.** The mean is at 2; 68% of the differences fall in the range −2 to +6, 95% in the range −6 to +10 and 99.7% in the range −10 to +14. See the Figure for Exercise 13.79a (below).

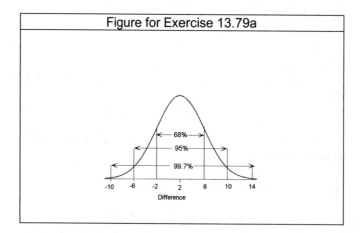

Figure for Exercise 13.79a

b. The null value of 0 falls half of a standard deviation below the mean of 2. Therefore, if the mean is really 2, it corresponds to an effect size of 0.5.

c. The power is .695, found from the row for $n = 20$ and the column for effect size of 0.5. The true effect size is found here as $d = \dfrac{\mu_1 - \mu_0}{\sigma} = \dfrac{2 - 0}{4} = 0.5$.

d. The power is the probability that the null hypothesis H_0: $\mu_d = 0$ will be rejected when in fact μ_d is 2 and the standard deviation of the differences is 4, for a one-sided test with $\alpha = .05$.

13.81 **a.** .6318 (from Minitab, Stat -> Power and sample size -> 1-sample t); use sample size = 45, "difference" = 0.3, standard deviation =1 and the option "greater than."

b. .9777 (from Minitab); use sample size = 30, "difference" = 13 − 10 = 3, standard deviation = 4 and the option "not equal."

c. .9789 (from Minitab); use sample size = 15, difference = −1.0, standard deviation = 1 and the option "less than."

13.83 The following values were found using software. Estimates based on the graph may differ somewhat from these values.

a. Power = .53 = the probability that the null hypothesis will be rejected for a one-sided t-test with $\alpha = .05$ when $n = 20$ and the true effect size is 0.4.

b. Power = .87 = the probability that the null hypothesis will be rejected for a one-sided t-test with $\alpha = .05$ when $n = 50$ and the true effect size is 0.4.

c. Power = .99 = the probability that the null hypothesis will be rejected for a one-sided t-test with $\alpha = .05$ when $n = 100$ and the true effect size is 0.4.

d. Power = .91 = the probability that the null hypothesis will be rejected for a one-sided t-test with $\alpha = .05$ when $n = 20$ and the true effect size is 0.7.

13.85 It is more useful to compare effect sizes because they reflect the magnitude of the true effect. The p-value is a function of the effect size and the sample size, so it is not a good way to compare whether the true effect is similar across studies when they have different sample sizes.

13.87 **a.** 0.231
b. 0.459
c. 0.280
d. No. In parts (a-c), $p − p_0 = 0.1$, but the effect size changes.
e. The effect size changes when p_0 changes, even if the difference $p − p_0$ stays fixed, because the denominator depends on p_0. So either the difference and p_0 must be specified or p and p_0 must be specified.

13.89 Issues 2, 4 and 5 should be of concern. As stated in #2, with a large sample size a statistically significant result may not have practical significance. Issue 4 is also concerned with this problem, and #5 helps to resolve it because the magnitude of the effect can be used to determine practical significance.

13.91 **a.** All except issue 5 rely on knowing the p-value. For #1, the item itself explains why. For #2, it would be useful to know if the result was just barely statistically significant, or if the p-value was extremely small. For #3 when the null hypothesis is not rejected it would help to know if the p-value indicated a result that was close to statistically significant or not. For #4 the same reasoning applies as in #2. For #6 it would be more convincing if the p-values for significant results were much less than .05.
b. Issues 2, 3 and 4. For #2, a confidence interval would allow us to see if the magnitude of the effect has real world importance. For #3, it would be useful to know if the confidence interval is centered on the null value, or if it is mostly above or below it. For #4 the reasoning is the same as for #2.
c. Issues 2, 3 and 4 all rely on sample size to determine whether or not they are problems. If the sample size is large and the test is statistically significant, the concerns in #2 and #4 should be taken into account. If the sample size is small and the test is not statistically significant, the concern in #3 should be taken into account.

13.93 You would want to know how many different relationships were examined. If this result was the only one that was statistically significant out of many examined, it could easily be a false positive. It could have occurred because of multiple testing.

13.95 **a.** Randomized experiment. The order of the treatments was randomly assigned.
b. Yes, a causal conclusion can be made because the study was a randomized experiment.
c. It could be single-blind, but not double-blind. The volunteers would have to know which drink they had before each session. The experimenter recording the length of time to exhaustion would not need to know which drink was consumed. The fact that the volunteers knew what they had consumed could be a problem because they may have altered their behavior to create the outcome they or the researchers wanted to see.

13.97 **a.** One-sided, H_a: $p > .5$ (majority thinks marijuana should be legalized).

b. Two-sided, H_a: $\mu \neq 500$ (mean is different from 500).

c. One-sided, H_a: $\mu_1 - \mu_2 < 0$, or equivalently, $\mu_1 < \mu_2$ (mean age of death is lower for left-handed people than for right-handed people).

d. Two-sided, H_a: $p_1 - p_2 \neq 0$, or equivalently, $p_1 \neq p_2$ (proportion different for male college students than for female college students).

13.99 **a.** Do not reject the null hypothesis. The p-value will be greater than $2(.111) = .222$ (based on information in Table A.3). An exact p-value, which can be determined using appropriate statistical software or the right calculator, is $2(.2631) = .5262$.

b. Reject the null hypothesis. For the new study, $t = 2.06$ and the p-value is less than $2(.024) = .048$. To determine the value of t, it helps to rewrite $t = \dfrac{\bar{x} - \mu_0}{s/\sqrt{n}}$ as $t = \dfrac{\sqrt{n}(\bar{x} - \mu_0)}{s}$.

The sample statistic $\bar{x}$, the null value (0), and the standard deviation are the same for both studies.

For the smaller study: $t = \dfrac{\sqrt{15}(\bar{x} - 0)}{s} = 0.65$

For the larger study: $t = \dfrac{\sqrt{150}(\bar{x} - 0)}{s} = \dfrac{\sqrt{10}}{1} \times \dfrac{\sqrt{15}(\bar{x} - 0)}{s} = \sqrt{10} \times 0.65 = 2.06$

$df = 150 - 1 = 149$. The exact p-value, which can be determined using appropriate statistical software or calculator, is $2(.0206) = .0412$.

c. The effect size was the same for both studies, but the conclusion was different because of the different sample sizes. The main point to be made is that sample size affects the statistical significance of a specific observed result. In this case, the same sample mean was not significantly different from 0 for a small sample, but was significantly different for the larger sample.

13.101 **a.** Step 1: H_0: $\mu_d = 0$

H_a: $\mu_d > 0$

μ_d = mean "husband age – wife age" difference in population of British married couples

Step 2: The sample size is sufficiently large to proceed. We must assume that the sample of married couples represents a random sample from the larger population of British married couples.

Test statistic is $t = \dfrac{\text{Sample statistic - Null value}}{\text{Null standard error}} = \dfrac{2.24 - 0}{0.3145} = 7.12$

Sample statistic = 2.24 years.

Null standard error $= \dfrac{s_d}{\sqrt{n}} = \dfrac{4.1}{\sqrt{170}} = 0.3145$ years.

Step 3: p-value ≈ 0. It is the area (probability) to the right of $t = 7.12$ in a t-distribution with $df = n - 1 = 170 - 1 = 169$. With Table A.3, it can be estimated from the $df = 100$ row that the p-value is less than .002. Because a t-distribution with a large value for df is similar to a standard normal curve, Table A.1 could be used to estimate that $P(t > 7.12)$ is less than .000000001.

Steps 4 and 5: Reject the null hypothesis using any reasonable value of α. The conclusion about the population of British married couples represented by this sample is that the mean difference between the ages of the husband and wife is greater than 0.

b. In part (a), the term "significant" refers to statistical significance, and in this situation, essentially means that we have concluded that the population mean difference in ages is not 0. However, the observed magnitude of the mean difference in ages, 2.24 years, does not have much practical importance, as it is not a large difference in ages. It could be misleading to say that British husbands are significantly older than their wives.

13.103 The maximum possible value of $p_1 - p_2$ is 1.0, so a one-sided 97.5% confidence interval is 0.02 to 1.0. Because the interval does not cover 0, we can reject H_0 using $\alpha = .025$

134

13.105 Reject the null hypothesis ($H_0: \mu_1 - \mu_2 = 0$) using $\alpha = .025$. The confidence interval does not cover 0. Because the alternative hypothesis is one-sided, the significance level corresponding to 95% confidence is $(1 - .95)/2 = .025$.

13.107 **a.** $H_0: \mu_1 - \mu_2 = 0$, equivalently, $\mu_1 = \mu_2$

$H_a: \mu_1 - \mu_2 < 0$, equivalently, $\mu_1 < \mu_2$ (lower mean estimate if Australia information given)

μ_1 = mean estimate for population of participants who would be told Australia's population

μ_2 = mean estimate for population of participants who would be told the U.S. population

b. Step 2: We must assume that the sample represents a random sample from a larger population. A comparative dotplot (or boxplot) shows an outlier within the responses from students who were told Australia's population. This is a possible violation of necessary conditions, but we will proceed because the point is a legitimate value (and it turns out not to distort the results). Note also that there is greater variation among responses from the U.S. group.

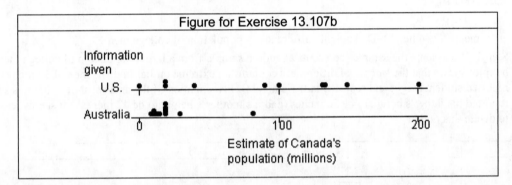

The appropriate test is a *t*-test for comparing two means. Minitab output for the unpooled version follows. Note that the standard deviations differ noticeably so the pooled version is inappropriate.

```
                    Output for Exercise 13.107
Two-sample T for Estimate

Informat      N        Mean        StDev     SE Mean
Australia     11       22.5        20.4         6.2
U.S.          10       88.4        66.1          21


Difference = mu (Australia) - mu (U.S. )
Estimate for difference:  -65.9
95% upper bound for difference: -26.4
T-Test of difference = 0 (vs <):
T-Value = -3.03   P-Value = 0.006   DF = 10
```

Step 2 continued and Steps 3, 4, and 5: The test statistic is $t = -3.03$ (with df = 10) and the *p*-value is .006. Reject the null hypothesis and conclude that in the population of students represented by this sample, the mean estimate of the population of Canada is lower if information about the population of Australia is provided than if information about the population of the U.S. is provided.

c. The outlier within the Australia group has not affected the results. That value pulls up the sample mean for the Australia group, but despite this the result is to declare the mean for this group is significantly less than for the U.S. group.

13.109 **a.** The confidence interval is Sample statistic $\pm\ t^* \times$ Standard error, where Sample statistic = $\bar{d} = 9.5$, $t^* = 2.26$ (from the df = 9 row and confidence level .95 column of Table A.2) and Standard error = $\dfrac{s_d}{\sqrt{n}} = \dfrac{3.6}{\sqrt{10}} = 1.14$. The interval is $9.5 \pm 2.26 \times 1.14$ or 9.5 ± 2.6 or 6.9 to 12.1 minutes.

b. Because 0 is not in the interval the null hypothesis can be rejected using $\alpha = .05$. Conclude that the population mean difference is not 0.

c. The interval is 6.9 to infinity.

d. Reject the null hypothesis using $\alpha = .025$, because 0 is not in the interval.

13.111 **a.** A type 1 error would mean that there is no change in mean cholesterol for the population but the test shows that there is. The consequence of a type 1 error is that patients are retested without cause because the levels on days 2 and 4 are not actually different. A type 2 error would mean that there is a drop in cholesterol on average from day 2 to day 4, but the test does not detect it. The consequence of a type 2 error is that patients should be retested but they will not be, so the initial cholesterol readings may be too high.

b. Type 2 seems more serious because patients will not be retested when they should be, and may end up taking medication that they do not need. With a type 1 error, patients would be retested when the retest is not needed, but there is no harm accompanying that error.

13.113 Step 1: H_0: $\mu_d \leq 0$

H_a: $\mu_d > 0$

μ_d = mean "Own height–Dad height" difference in population of college men

Step 2: We assume the sample represents a random sample from a larger population of college men. A boxplot shows that the sample of differences contains an extreme outlier (a difference of 37 inches between reported student height and reported father's height). Investigation of the data set shows that this student reported his father's height to be 32 inches (and his mother's height to be 42 inches). It seems reasonable to delete this outlier.

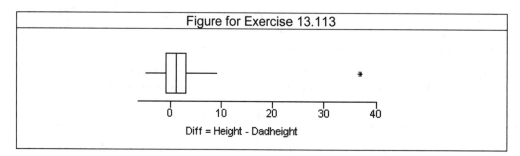

Figure for Exercise 13.113

Diff = Height - Dadheight

The appropriate procedure is a t-test for paired data. Minitab output is given below. The outlier was deleted.

```
Output for Exercise 13.113 (outlier deleted)
Paired T for Height - dadheight

                  N       Mean     StDev    SE Mean
Height           75     70.143     3.089     0.357
dadheight        75     69.060     3.691     0.426
Difference       75      1.083     3.010     0.348

95% lower bound for mean difference: 0.504
T-Test of mean difference = 0 (vs > 0):
T-Value = 3.12   P-Value = 0.001
```

Step 2 continued and Steps 3, 4, and 5: The test statistic is $t = 3.12$ and the p-value is .001. Reject the null hypothesis and conclude that in the population of college men, students' heights are greater, on average than fathers' heights. The observed magnitude of the difference is $\bar{d} = 1.083$ inches.

13.115 Step 1: H_0: $\mu_1 - \mu_2 = 0$, or equivalently, $\mu_1 = \mu_2$

H_a: $\mu_1 - \mu_2 > 0$, or equivalently, $\mu_1 > \mu_2$

μ_1 = mean hours of sleep for population of students who do not feel sleep deprived.

136

μ_2 = mean hours of sleep for population of students who do feel sleep deprived.

<u>Step 2</u>: We assume the students in the sample are representative of the population of all college students. The sample sizes are sufficiently large to proceed, although a graph of the data still should be done. Comparative boxplots are shown below. There is an outlier at 0 hours in the deprived = "yes" group (see the boxplot below). If that outlier is removed, the "yes" mean changes to 6.06 hours and the difference is 1 hour. The outlier is clearly a mistake. Students were asked how many hours they *usually* sleep per night, and if someone actually slept 0 hours per night they wouldn't be around to talk about it! So it is reasonable to remove the outlier. The analysis here is shown both with and without the outlier.

Figure for Exercise 13.115

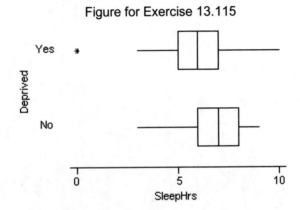

Minitab output for the unpooled version follows.

Output for Exercise 13.115 (all data)

```
Two-sample T for SleepHrs

Deprived   N   Mean   StDev   SE Mean
No        35   7.06   1.16     0.20
Yes       51   5.94   1.35     0.19


Difference = mu (No) - mu (Yes)
Estimate for difference:  1.11597
95% lower bound for difference:  0.66267
T-Test of difference = 0 (vs >): T-Value = 4.10   P-Value = 0.000   DF = 79
```

Output for Exercise 13.115 (outlier removed)

```
Two-sample T for SleepHrs

Deprived   N   Mean   StDev   SE Mean
No        35   7.06   1.16     0.20
Yes       50   6.06   1.06     0.15


Difference = mu (No) - mu (Yes)
Estimate for difference:  0.997143
95% lower bound for difference:  0.585503
T-Test of difference = 0 (vs >): T-Value = 4.04   P-Value = 0.000   DF = 68
```

<u>Step 2 continued and Steps 3, 4, and 5 (all data)</u>: The test statistic is $t = 4.10$ (with df = 79) and the *p*-value is .000. Reject the null hypothesis and conclude that in the population of students represented by this sample, students who feel sleep-deprived do in fact get less sleep on average than does the general population. The observed difference in sample means is $\bar{x}_1 - \bar{x}_2 = 7.06 - 5.94 = 1.12$ hours.

<u>Step 2 continued and Steps 3, 4, and 5 (outlier removed)</u>: The test statistic is $t = 4.04$ (with df = 68) and the *p*-value is .000. Reject the null hypothesis and conclude that in the population of students represented by this sample, students who feel sleep-deprived do in fact get less sleep on average than does the general population. The observed difference in sample means is $\bar{x}_1 - \bar{x}_2 = 7.06 - 6.06 = 1.0$ hour.

137

CHAPTER 14
ODD-NUMBERED SOLUTIONS

14.1   **a.** $\hat{y} = 119 - 1.64(40) = 119 - 65.6 = 53.4$ degrees Fahrenheit.

        b. $e_i = y_i - \hat{y}_i = 50 - 53.4 = -3.4$ degrees.

14.3   **a.** β_0 = intercept of the regression line in the population. β_1 = slope of regression line in the population and is the average increase in y per each one unit increase in x.

        b. $E(y_i) = \beta_0 + \beta_1 x_i$

14.5   **a.** Slope = 0.9. Mean systolic blood pressure increases 0.9 points per each 1-year increase in age.
        b. Predicted systolic blood pressure = 85 + 0.9 (50) = 130.
        c. Residual = Actual – Predicted = 125 – 130 = –5.
        d. The man's blood pressure is 5 points lower than the predicted blood pressure for someone his age.

14.7   **a.** Slope = 2.90. Mean desired height increases 2.90 pounds per each 1-inch increase in height.
        b. Correct notation is b_1. This is a slope for a sample.
        c. Predicted desired weight = –65.4 + 2.90 (65) = 123.1 pounds for women who are 65 inches tall.

14.9   **a.** $x = 70$.
        b. $y = 180$.
        c. $E(Y) = -250 + 6(70) = 170$.
        d. Deviation = 180 – 170 = 10.
        e. The component that is explained by knowing x is 170.
        f. The unexplained part of y is the deviation, 10.

14.11   **a.** $R^2 = \dfrac{\text{SSTO} - \text{SSE}}{\text{SSTO}} = \dfrac{500 - 300}{500} = 0.40$, or 40% when converted to a percent.

        b. $R^2 = \dfrac{200 - 40}{400} = 0.80$, or 80%.

        c. $R^2 = \dfrac{80 - 0}{80} = 1.00$, or 100%. Note that $R^2 = 100\%$ whenever SSE = 0.

        d. $R^2 = \dfrac{100 - 95}{100} = 0.05$, or 5%.

14.13   **a.** $s = 2.837$. It is roughly, for any specific latitude, the average (absolute) difference between actual April temperatures and the predicted temperature for that latitude.

        b. Geographic latitude explains 91.7% of the variation among April temperatures for the cities in the sample.

14.15   **a.** $\hat{y}_i = 577 - 3.01(21) = 513.79$ ft. (about 514 ft.)
        b. $e_i = y_i - \hat{y}_i = 525 - 513.79 = 11.21$ ft. (about 11 ft.)

        c. $513.79 \pm (2 \times 50)$ which is 413.79 ft. to 613.79 ft. (about 414 to 614 ft.). From the Empirical Rule about 95% of individual values are within two standard deviations of the mean.
        d. Yes. 650 feet is more than 2 standard deviations from the mean distance for drivers who are 21 years old, so it would be unusual. Notice that 650 feet is outside the interval calculated in the previous part.

139

14.17 **a.** For each observation, first calculate $\hat{y}$ and then calculate $e = y - \hat{y}$. The residuals are shown in the last row of the following table.

x	1	1	3	3	5	5
y	10	12	13	17	17	21
$\hat{y} = 9 + 2x$	11	11	15	15	19	19
$e = y - \hat{y}$	−1	1	−2	2	−2	2

The sum of residuals is $\sum e_i = -1 + 1 - 2 + 2 - 2 + 2 = 0$

b. $SSE = \sum e_i^2 = (-1)^2 + 1^2 + (-2)^2 + 2^2 + (-2)^2 + 2^2 = 18$

c. $s = \sqrt{\dfrac{SSE}{n-2}} = \sqrt{\dfrac{18}{6-2}} = 2.12$

14.19 **a.** SSTO = 1746.8, found in the "Total" row and "SS" column.

b. SSE = 144.9, found in the "Residual Error" row and "SS" column.

c. $r^2 = \dfrac{SSTO - SSE}{SSTO} = \dfrac{1601.9}{1746.8} = .917$ or 91.7%.

14.21 **a.** No. The p-value is not less than 0.05, the usual standard for statistical significance.

b. The p-value would be 0.552, the same as it is for the test of whether the slope is 0.

14.23 **a.** H_0: $\beta_1 = 0$ versus H_a: $\beta_1 \neq 0$, where β_1 = slope of the regression equation.

b. Reject the null hypothesis; conclude that there is a linear relationship. The value of the t-statistic is −14.11 and the p-value is 0.000.

c. $t = \dfrac{-1.6436 - 0}{0.1165} = -14.11$

d. The same as for the test in part (a), which is 0.000. The correlation is 0 if and only if the slope is 0.

14.25 **a.** The relationship is not statistically significant. The p-value is .165 for the significance test of H_0: $\beta_1 = 0$ versus H_a: $\beta_1 \neq 0$.
b. The p-value is the same as it is for part (a) so it is .165 (which means we cannot reject the null hypothesis).

14.27 **a.** The notation is b_1 because 7 is the slope for the sample regression line.

b. The approximate 95% C.I. is $7 \pm (2 \times 1.581)$ which is 7 ± 3.162, or about 3.84 to 10.16. With 95% confidence, we can say that in the population of college men, the mean increase in weight per 1-inch increase in height is somewhere between about 3.84 and 10.16 pounds.
c. The confidence interval for the slope contains 5, so the statement about the population slope is reasonable.

14.29 **a.** There are two sets of hypotheses being tested. In each case the hypotheses are H_0: $\rho = 0$ versus Ha: $\rho \neq 0$. (It may be that in reality the researchers' alternative hypothesis was $\rho > 0$, but computer software routinely provides a p-value for the two-sided alternative.) In the first set (with $p < 0.001$), ρ is the population correlation between weekly time spent watching music videos and concern about weight. In the second set (with $p < 0.05$), ρ is the population correlation between weekly time spent watching music videos and importance of appearance.
b. The researchers have decided to reject the null hypothesis in both cases, quoting p-values less than .001 and less than .05.

140

14.31 **a.** Prediction interval for a value of y because the interest is in the GPA of one student.
b. Confidence interval for the mean because the interest is in the mean for a population with a certain high school GPA.

14.33 **a.** Grade point average of an individual student who typically misses two classes per week.
b. Mean grade point average for all students who typically miss two classes per month.

14.35 **a.** 78.02 to 82.67. With 95% confidence, we can say that the mean pulse rate after marching is between 78.02 and 82.67 for those in the population whose pulse is 70 before marching.

b. 65.50 to 95.19. In a population of individuals with a pulse rate of 70 before marching, about 95% of the individuals will have a pulse between 65.50 and 95.19 after marching.

14.37 **a.** For $x = 40$, $\hat{y}_i = 3.59 + 0.9667(40) = 42.258$.

b. The prediction interval is wide because there is variation among the individual ages of husbands married to women 40 years old. The confidence interval is narrow because, with $n=170$ the regression line (and hence the mean) can be estimated with good precision.

14.39 The calculation is $42.258 \pm (2 \times 0.313)$, which gives approximately the interval provided by Minitab. The general format is *estimate $\pm$ 2 standard error*. The *estimate* is the value under "Fit" and the relevant *standard error* is value under "SE Fit". A slight discrepancy occurs because Minitab used $t^* = 1.974$ as the "exact" multiplier (rather than 2). The degrees of freedom for the t-distribution used to determine the exact multiplier are df $= n-2 = 170-2 = 168$.

14.41 **a.** Conditions 2 (no outliers) and 4 (normal distribution). A histogram of residuals can be used to determine if there are outliers and can be used to judge the shape of the distribution.
b. Conditions 1 (linear relationship), 2 (no outliers), and 3 (constant variance). A scatterplot of residuals versus x can be used to judge whether the right form of regression equation was used, whether there are any outliers, and whether the standard deviation is about the same for all values of x.

c. Condition 5 (independence).

d. Conditions 1, 2, and 3. A scatterplot of y versus x can be used to judge whether the right form of regression equation was used, whether there are any outliers, and whether the standard deviation is about the same for all values of x.

14.43 There is an outlier (at about 14 as a value of the residual). Aside from the outlier, the distribution might be normal although it is difficult to judge because of the relatively small sample size.

14.45 **a.** Condition 2 is violated because there appears to be an outlier (the observation at *Neck girth* $\approx$ 10 with *Weight* $\approx$ 150).
b. The outlier is so inconsistent with the remainder of the data set that it is almost certainly a mistake. It might be dropped from the data set. It seems far too inconsistent with the other points. If possible, check the source of the data to see if a mistake has been made.
c. The weights for most data points seem to be within 40 pounds (one way or the other) from the line, so most of the frequency in the histogram will be in this range. For the outlier, the weight is more than 100 pounds from the line so in the histogram, there'll be a bar set apart from the other bars. Here's the actual histogram of residuals for this data set: Figure for this exercise is on the next page.

Figure for Exercise 14.45c

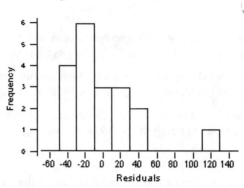

14.47 **a.** This is given by the sample slope $b_1 = -0.269$. (Or, with more decimal places, $b_1 = -0.26917$.)
b. Approximately $b_1 \pm 2\,s.e.(b_1)$. This is $-0.26917 \pm (2 \times 0.6616)$ or about -0.401 to -0.137. With 95% confidence, we can say that the slope of the regression line in the population is between -0.401 and -0.137. *Note*: The exact multiplier in this situation is $t^* = 1.981$ which can be found by using software like Minitab or Excel to find the value of t^* for which the probability is 0.95 in a *t*-distribution that has df $= n - 2 = 116 - 2 = 114$.
c. $H_0: \beta_1 = 0$ versus $H_a: \beta_1 \neq 0$.
d. $t = \dfrac{b_1}{s.e.(b_1)} = \dfrac{-0.26917}{0.06616} = -4.07$, df $= n - 2 = 116 - 2 = 114$.
e. The relationship is statistically significant. The *p*-value is 0.000, given in the output under "*P*" in the row labeled *study*.

14.49 $b_0 = 7.56$ hours. Yes, the intercept has a logical interpretation here. It is the estimated mean hours of sleep for students who studied 0 hours. As Figure 3.14 (page 85) shows, 0 hours is part of the range of the data.

14.51 The prediction interval is 3.746 to 9.750 hours of sleep. The two interpretations are: (1) Of students in the population who studied 3 hours, about 95% slept between 3.746 and 9.75 hours. (2) The probability is .95 that the hours of sleep will be in the given interval for a student randomly selected from the population of students who studied 3 hours.

14.53 **a.** There is a positive association and the relationship might be linear although it's difficult to be certain because there are so few observations and there may be outliers. Figure for this exercise is on the next page.

142

Figure for Exercise 14.53a

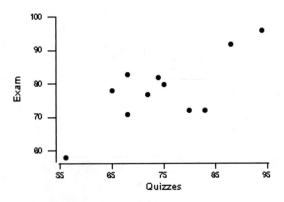

b. Minitab gives the following output:

Output for Exercise 14.53b				
The regression equation is				
Exam = 25.4 + 0.707 Quizzes				
Predictor	Coef	SE Coef	T	P
Constant	25.38	16.22	1.57	0.152
Quizzes	0.7069	0.2147	3.29	0.009

c. For quiz average equal to 75, $\hat{y} = 25.38 + 0.7069(75) = 78.4$

d. Minitab gives the 50% prediction interval as 72.97 to 83.83 (approximately 73 to 84). This interval estimates the central 50% of exam scores for students with a quiz average equal to 75.

14.55 **a.** R^2=44.7%. Father's height explains about 44.7% of the variation in son's height.

b. $r = \sqrt{0.447} = 0.67$. The correlation is the square root of R^2 (expressed as a decimal fraction). The sign of the correlation is the same as the sign of the slope, which is positive in this example.

14.57 *Step 1*: The hypotheses are H_0: β_1=0 versus H_a: $\beta_1 \neq 0$.
Step 2: To check the necessary conditions described in section 14.5, examine residual plots as well as a scatter plot of y versus x. The test statistic is $t = \dfrac{b_1}{s.e.(b_1)}$ and statistical software will provide the value.

Step 3: Statistical software will provide the *p*-value. A t-distribution with df = n–2 is used to find this value.
Step 4: Reject the null hypothesis if the p-value is less than the specified significance level (usually α=.05).
Step 5: Write a conclusion that describes whether the relationship between the explanatory and response variables is statistically significant or not.

14.59 **a.** There is a positive association with a linear pattern. Figure for this exercise is on the next page.

143

Figure for Exercise 14.59a

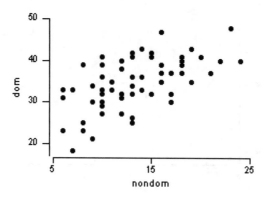

b. The equation is $\hat{y} = 22.693 + 0.876x$.

Output for Exercise 14.59b				
The regression equation is				
dom = 22.7 + 0.876 nondom				
Predictor	Coef	SE Coef	T	P
Constant	22.693	2.079	10.92	0.000
nondom	0.8760	0.1506	5.82	0.000
S = 5.051	R-Sq = 35.7%		R-Sq(adj) = 34.6%	

c. The standard deviation is $s = 5.051$ and $R^2 = 35.7\%$. On average, the actual number of letters written with the dominant hand differs by about 5.051 letters from the regression line prediction based on the non-dominant hand. Knowledge of the number of letters written with the non-dominant hand explains 35.7% of the observed variation in letters written with the dominant hand.

d. An approximate 95% interval is $b_1 \pm 2\,s.e.(b_1)$ which is $0.876 \pm (2 \times 0.1506)$ which is about 0.575 to 1.177. With 95% confidence, we can say that the slope of the regression line in the population is somewhere in this interval. This slope, by the way, estimates the increase in mean letters written with the dominant hand for each 1-letter increase in letters written with the non-dominant hand.

e. An equation for the statement is *dom* = 23 + *nondom*, an equation for which the intercept is 23 and the slope is 1. Notice that the intercept is 22.7 for the sample regression line and that the value 1 is contained within the interval determined in part (d) for the population slope. So, the statement is reasonable.

14.61 **a.** There is a strong positive association between weight and chest girth for these bears. The two observations with the greatest chest girths could be outliers, or there may be a gently curving pattern.

Figure for Exercise 14.61a

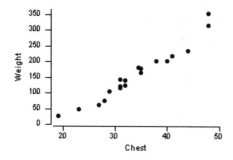

b. The regression equation is $\hat{y} = -206.9 + 10.76x$. Minitab output (for parts b to f) is on the next page.

```
                     Output for Exercise 14.61b
The regression equation is
Weight = - 207 + 10.8 Chest

Predictor        Coef     SE Coef          T          P
Constant      -206.90       20.69     -10.00      0.000
Chest         10.7595      0.5937      18.12      0.000


S = 19.57        R-Sq = 95.1%      R-Sq(adj) = 94.8%


New Obs    Fit    SE Fit      95.0% CI            95.0% PI
1       223.48      5.72   (211.41, 235.55)   (180.47,   266.49)
```

c. $R^2 = 95.1\%$. The variable chest girth explains 95.1% of the observed variation in weight.

d. $\hat{y} = -206.9 + 10.76(40) = 223.5$ lbs. Notice that Minitab gives $\hat{y} = 223.48$. The program kept track of more decimal places than we did.

e. The 95% prediction interval is 180.47 lbs. to 266.49 lbs. About 95% of bears with chest girth equal to 40 inches will weigh between 180.47 pounds and 266.49 pounds.
f. The 95% confidence interval for the mean is 211.41 lbs. to 235.55 lbs. We are 95% confident that for bears with a chest girth equal to 40 inches the mean weight is between 211.41 pounds and 235.55 pounds.

14.63   **a.** With all data points included in the analysis, Minitab gives the 90% prediction interval as 67.050 inches to 77.731 inches.
b. The interval from part (c) of the previous exercise was 68.517 inches to 75.512 inches. The interval computed in part (a) of this problem is wider because the estimated standard deviation from the regression line is larger than it is in the previous exercise due to the inclusion of the outlier. The inflated standard deviation creates a wider interval describing the variation among individual heights. This makes sense, because with the outlier included, there is indeed more variability among the heights in the sample.
c. The residual plot clearly shows that there is an outlier in the data set.

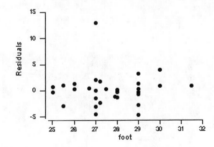

Figure for Exercise 14.63c

CHAPTER 15
ODD-NUMBERED SOLUTIONS

15.1   **a.** Appropriate. Both variables are categorical.
b. Not appropriate. Both variables are quantitative.

15.3   **a.** Null: Exam cheating is not related to feelings about importance of religion.
Alternative: Exam cheating is related to feelings about importance of religion.
b. The p-value is greater than .05, so we are not able to reject the null hypothesis. We cannot conclude that the two variables are related.

c. Degrees of freedom = $(3 - 1)(2 - 1) = 2$.

15.5   **a.** p-value $=.05$
b. $.10 < p$-value $< .25$ (based on Table A.5). With Excel, p-value = CHIDIST(6.7,4) =.1526.
c. p-value $<.001$ (based on Table A.5). With Excel, CHIDIST(26.23,2) =.000002.

d. p-value $>.50$ (based on Table A.5). With Excel, CHIDIST(2.28,9) =.986

15.7   **a.** 3.84
b. 16.81 [df =(3–1) (4–1) =6]
c. 12.59

15.9   **a.** H_0: Gender and type of class taken are not related for the population of students.
H_a: Gender and type of class taken are related for the population of students.
b. df=1.
c. 1.258
d. p-value =.262. Using Table A.5, p-value is between .25 and .50.
e. Do not reject the null hypothesis for α = .05. The observed relationship between gender and type of class taken is not statistically significant.

15.11   **a.** No, there is no obvious choice. Both variables are opinions and neither precedes the other.
b. Row percentages: Of those in favor of the death penalty, 40.44% think marijuana should be legal and 59.56% think it should not be legal. Of those opposed to the death penalty, 40.05% think marijuana should be legal and 59.95% think it should not be legal.
Column percentages: Of those thinking marijuana should be legal, 65.48% favor the death penalty and 34.52% are opposed. Of those thinking marijuana should not be legal, 65.11% favor the death penalty and 34.89% are opposed.
Both sets of conditional percentages are evidence that there may not be a relationship. For instance, the percentage favoring legalization of marijuana is about the same for those opposed to the death penalty as is for those in favor of the death penalty.
c. H_0: Opinion about death penalty and opinion about marijuana legalization are not related.
H_a: Opinion about death penalty and opinion about marijuana legalization are related.

d. Minitab output is given below (expected counts are beneath observed counts in each cell). χ^2 =0.017, df = 1, p-value = 0.896 If we use Table A.5, p-value >.50.
We cannot reject the null hypothesis. We are ***not*** able to conclude that there is a relationship between opinion about the death penalty and opinion about marijuana legalization.

Output for Exercise 15.11d			
	Legal	NotLegal	All
Favor	313	461	774
	311.9	462.1	774.0

```
Oppose        165        247        412
             166.1       245.9     412.0

All           478        708       1186
             478.0       708.0    1186.0

Cell Contents:        Count
                      Expected count

Pearson Chi-Square = 0.017, DF = 1, P-Value = 0.896
```

15.13 For expected counts, proportion with an ear infection is .22 within each treatment.
Placebo: 39.07/178 = .22
Xylitol gum: 39.29/179 = .22
Xylitol lozenge: 38.63/176 = .22

15.15 Step 1: H_0: Typical grades and seat belt use are not related for the population of 12th graders.
H_a: Typical grades and seat belt use are related for the population of 12th graders.
Step 2: Expected counts are all greater than 5 so proceed with the chi-square test. Minitab output is shown below. Test statistic is $\chi^2 = 77.776$, df = 2.

Steps 3, 4, and 5: p-value =.000.

Reject the null hypothesis.

Typical grades and seat belt use are related for the population of 12th graders.

```
                    Output for Exercise 15.15
Expected counts are printed below observed counts

         Most Times  Rarely   Total
A or B      1354      180      1534
           1292.56   241.44

   C         428      125       553
            465.96    87.04

D or F        65       40       105
             88.47    16.53

Total       1847      345      2192

Chi-Sq =  2.920 + 15.634 +
          3.093 + 16.558 +
          6.228 + 33.343 = 77.776
  DF = 2, P-Value = 0.000
```

15.17 **a.** Row percents are:

	Frequency of Reading Newspapers				
Age	*Every Day*	*A Few Times a Week*	*Once a Week*	*Less Than Once a Week*	*Total*
18-29	19.2%	29.1%	16.2%	35.5%	100%
30-49	22.8%	24.1%	19.3%	33.8%	100%
≥ 50	46.9%	17.9%	12.5%	22.7%	100%

Those in the oldest age group are much more likely to read a newspaper everyday than those in the younger two age groups. There's not much difference between the two younger age groups.

148

b. H$_0$: Age and frequency of reading the newspaper are not related

H$_a$: Age and frequency of reading the newspaper are related

c. $\chi^2 = 96.488$, df = (3 -1)(4-1) = 6, p-value ≈ 0. We can reject the null and infer that age and frequency of reading the newspaper are related.

Minitab output follows. Expected counts are shown below observed counts.

```
                        Output for Exercise 15.17c

                   Few times   Once a  Less than
             Every day  a week    week  once/week    All

18-29              45       68      38         83    234
                 75.8     52.3    37.1       68.8  234.0

30-49             118      125     100        175    518
                167.8    115.8    82.1      152.3  518.0

50+               260       99      69        126    554
                179.4    123.9    87.8      162.9  554.0

All               423      292     207        384   1306
                423.0    292.0   207.0      384.0 1306.0


Cell Contents:       Count
                     Expected count

Pearson Chi-Square = 96.488, DF = 6, P-Value = 0.000
```

As an example of an expected count, for the 18-29, Every day cell, the expected count is $\dfrac{234 \times 423}{1306} = 75.8$.

15.19 $\chi^2 = \dfrac{N(AD - BC)^2}{R_1 R_2 C_1 C_2} = \dfrac{1186(313 \times 247 - 478 \times 708)^2}{774 \times 412 \times 478 \times 708} = 0.017$.

15.21 **a.** The researchers probably thought, in advance of collecting data, that short students would be more likely to be bullied. So, a one-sided alternative would be appropriate and a z-test for comparing two proportions should be used.

b. Step 1: H$_0$: $p_1 - p_2 = 0$ (or $p_1 = p_2$) versus Ha: $p_1 - p_2 > 0$ (or $p_1 > p_2$)

p_1 = proportion ever bullied in population of short students

p_2 = proportion ever bullied in population of students not short

Step 2: Sample sizes are sufficiently large and we assume the samples represent random samples.

The test statistic is $z = 3.02$. Output for comparing two proportions is given below. Alternatively, a chi-square test can be done and the relationship $z = \sqrt{\chi^2}$ used to determine the z-statistic.

Steps 3, 4, and 5: p-value =.001 (reported in output). Reject the null hypothesis, and conclude that a higher proportion of short students than non-short students have been bullied.

```
                Output for Exercise 15.21


Sample       X       N  Sample p
1           42      92  0.456522  (short)
2           30     117  0.256410  (not short)

Estimate for p(1) - p(2):   0.200111
```

```
95% lower bound for p(1) - p(2):  0.0919200
Test for p(1) - p(2) = 0 (vs > 0):  Z = 3.02  P-Value = 0.001
```

c. For short students, the 95% C.I. is about .36 to 56. For students not short, the C.I is about .18 to .35. To do "by hand" calculations use $\hat{p} \pm 2\sqrt{\dfrac{\hat{p}(1-\hat{p})}{n}}$. Short students: $\hat{p} = .457$ and n = 92. Not short: $\hat{p} = .256$ and $n = 117$. The confidence intervals do not overlap, further confirming the result in part (b) that the population proportion is higher for the short students.

15.23 No. The response variable has 3 categories.

15.25 **a.** Given that 2 out of 10 participants have reduced pain, what is the probability that both of them would be in the magnet-treated group?

b. The null hypothesis is that there is no relationship between treatment type and pain reduction, while the alterative hypothesis is that the magnet treatment is more likely to reduce pain. The one-tailed p-value, which is appropriate in this case, is .222. Do not reject the null hypothesis.

15.27 **a.** H_0: Seal performance and temperature are unrelated.

H_a: Seal performance depends on temperature.

b. H_0: Population proportions planning to vote for each candidate were the same in September and June.

H_a: Population proportions planning to vote for each candidate are differed in September and June.

c. H_0: Taking a typing class does not reduce the chance of getting carpal tunnel syndrome.

H_a: Taking a typing class does reduce the chance of getting carpal tunnel syndrome.

15.29 As the sample size increases a fixed amount of difference becomes more significant. (See page 598 in the text.)

15.31 **a.** (190 + 26)/400 = .54; combine the two categories where X was preferred before

b. 26 + 14 = 40; combine the last two categories given in the table

c. H_0: $p \geq .5$ versus Ha: $p < .5$

d. p-value = .0403. This is the probability less than or equal to 14 in a binomial distribution with $n = 40$ and $p = .5$. In Excel, the p-value can be found using BINOMDIST(14, 40,.5,1).

The p-value is less than .05 (level of significance), so we reject the null and infer that the proportion preferring X decreased following the debate.

15.33 **a.** 100 for each of the 3 categories, calculated as 300(1/3) = 100.

b. 250, 250 and 500, respectively, calculated as 1000(1/4), 1000(1/4) and 1000(1/2).

15.35 **a.** No. A chi-square statistic is a sum of non-negative numbers so it would have to be greater than or equal to 0.

b. Yes. This would happen if the observed count equaled the expected count in each cell of the table.

c. Yes. For example, see the solution for exercise 15.32.

d. No. The observed counts will always be whole numbers because they are actual counts.

e. No. The null hypothesis gives hypothesized probabilities for all possible categories so the sum of probabilities must be 1.

f. No. For a goodness of fit test, df = number of categories – 1.

15.37 **a.** H_0: p = $\frac{1}{10}$ for each of the 10 numbers, where p = the probability of a student choosing that number. The alternative hypothesis is that not all of the probabilities are $\frac{1}{10}$.

b. The calculations are as follows, using the expected count of 19 for each category:

$$\chi^2 = \frac{(2-19)^2}{19} + \frac{(9-19)^2}{19} + \frac{(22-19)^2}{19} + \frac{(21-19)^2}{19} + \frac{(18-19)^2}{19} +$$
$$\frac{(23-19)^2}{19} + \frac{(56-19)^2}{19} + \frac{(19-19)^2}{19} + \frac{(14-19)^2}{19} + \frac{(6-19)^2}{19} = 104.32.$$

c. Degrees of freedom = k – 1 = 10 – 1 = 9.

d. The *p*-value is essentially 0. From Table A.5, p-value < .001 because 104.32 > 27.88.

e. Using any reasonable level of significance the conclusion is the same. At least two of the probabilities are not 1/10. Students are not equally likely to choose each of the digits.

15.39 **a.** Possible sequences are HH, TH, HT, TT.
Probabilities are: X = 0, p_0 = 1/4 = .25; X = 1, p_1 = 1/2 = .5; X = 2, p_2 = 1/4 = .25
b. H_0: p_0 =.25, p_1 =.50, p_2 =.25
H_a: not all p_i are as specified in H_0

c. Step 2: Expected counts: X = 0, 60×.25 = 15; X = 1, 60×.5 = 30; X = 2, 60×.25 = 15.

All expected counts are greater than 5 so proceed with the chi-square test.

Test statistic is $\chi^2 = \frac{(8-15)^2}{15} + \frac{(40-30)^2}{30} + \frac{(12-15)^2}{15} = 7.2$; df = *k*–1 = 3–1 =2.

Steps 3, 4, and 5: .025< *p*-value < .05 (based in Table A.5). With Excel, *p*-value = CHIDIST(7.2,2) = .027.

Reject the null hypothesis. Conclude that observed student results are significantly different from expected results based on theory.

d. Exactly one head occurred more often than expected. Perhaps some students made up their data, and may have had a tendency to claim they got 1 head in the 2 flips.

15.41 Step 1: H_0: p_1 =.5, p_2 =.3, p_3 =.2 (manufacturer's hypothesis)
H_a: not all p_i are as specified in H_0
Step 2: Expected counts: Silver, 111×.5 = 55.5; Blue, 111×.3 = 33.3; Green, 111×.2 = 22.2.

All expected counts are greater than 5 so proceed with the chi-square test.

Test statistic is $\chi^2 = \frac{(59-55.5)^2}{55.5} + \frac{(27-33.3)^2}{33.3} + \frac{(25-22.2)^2}{22.2} = 1.766$; df= *k*–1= 3–1= 2
Steps 3, 4, and 5: .25 < *p*-value < .50 (based on Table A.5).
With Excel, *p*-value = CHIDIST(1.766,2) = .4135.
Do not reject the null hypothesis. Based on this sample, the manufacturer's hypothesis about color preferences is not rejected.

15.43 Step 1: H_0: p_1 = .3, p_2 = .2, p_3 = .2, p_4 = .1, p_5 = .1, p_6 = .1 (manufacturer's claim)
H_a: not all p_i are as specified in H_0
Step 2: Expected counts: Brown: 2081×.3 = 624.3; Red and Yellow, 2081×.2 = 416.2; Blue, Orange and Green, 2081×.1 = 208.1.

All expected counts are greater than 5 so proceed with the chi-square test.

Test statistic is $\chi^2 = \dfrac{(602-624.3)^2}{624.3} + \dfrac{(396-416.2)^2}{416.2} + \dfrac{(379-416.2)^2}{416.2} +$

$\dfrac{(227-208.1)^2}{208.1} + \dfrac{(242-208.1)^2}{208.1} + \dfrac{(235-208.1)^2}{208.1}$

$= 15.8$; df $= k-1 = 6 - 1 = 5$.

<u>Steps 3, 4, and 5</u> $.005 < p\text{-value} < .01$; using Excel, CHIDIST(15.8,5) = .007. Reject the null hypothesis and conclude that proportions are not as stated at M&M web site for the population from which these bags were sampled.

15.45 **a** The age groups are not markedly different. Women in the youngest age group are somewhat less likely to be dissatisfied with their appearance.

Age	Satisfaction with Appearance			
	Very	Somewhat	Not Too	Not at All
Under 30	31.9%	58.2%	7.1%	2.8%
30 – 49	24.8%	57.1%	16.0%	2.0%
Over 50	34.0%	49.0%	13.1%	3.9%

Figure for Exercise 15.45a

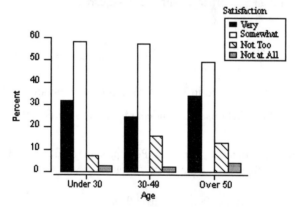

b. df $= (r-1)(c-1) = (3-1)(4-1) = 6$

c. Yes, the relationship is statistically significant (p-value is reported to be .027).

d. The p-value is the area to the right of 14.278 in a chi-square distribution with df $= 6$. The p-value represents the probability of observing a relationship as strong as or stronger than the one observed in the sample, if there really is no relationship in the population.

Figure for Exercise 15.45d

152

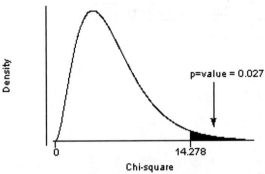

p=value = 0.027

14.278

Chi-square

e. Expected Count $= \dfrac{\text{Row Total} \times \text{Column Total}}{\text{Total n}} = \dfrac{312 \times 98}{747} = 40.93$

15.47 **a.** There appears to be a relationship. As the number of ear pierces increases, the percentage with a tattoo also increases.

Pierces	% with tattoo
2 or less	7.4% (40/538)
3 or 4	13.4% (58/432)
5 or 6	27.6% (77/279)
7 or more	42.1% (53/126)

The relationship can also be described using the column percentage. In the following bar graph, we see that women who have a tattoo tend to have more ear pierces than women who don't have a tattoo.

Figure for Exercise 15.47a

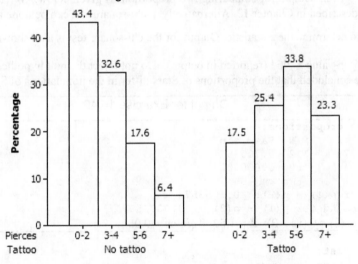

b. H_0: Number of ear pierces and having a tattoo (or not) are not related
  H_a: Number of ear pierces and having a tattoo (or not) are related

c. Minitab output is given on the next page (expected counts are beneath observed counts in each cell).

$\chi^2 = 119.279$, df = 3, p-value =.000. If we use Table A.5, p-value <.001.

Reject the null hypothesis. Conclude that for the population of college women represented by the sample, there is a relationship between number of ear pierces and having a tattoo or not.

153

```
                        Output for Exercise 15.47c
                No      Yes    Total
        0-2     498      40     538
                448.79   89.21

        3-4     374      58     432
                360.37   71.63

        5-6     202      77     279
                232.74   46.26

        7+       73      53     126
                105.11   20.89

   Total        1147    228    1375

   Chi-Sq = 119.279, DF = 3, P-Value = 0.000
```

15.49 **a.** Sheep performed slightly better. For the Sheep, 70.5% were Stars and 29.5% were Duds. For the Goats, 60% were Stars and 40% were Duds. An appropriate graph is a bar graph displaying these percentages.
b. The test should be one-sided because the speculation before collecting the data was that Sheep will perform better than Goats. Therefore the hypotheses are:

$$H_0: p_1 - p_2 = 0 \text{ (or } p_1 = p_2 \text{) versus } H_a: p_1 - p_2 > 0 \text{ (or } p_1 > p_2 \text{)}$$

p_1 = proportion of Stars in the population of students who are Sheep

p_2 = proportion of Stars in the population of students who are Goats

c. Step 2: Sample sizes are sufficiently large and we assume the samples represent random samples. The test statistic is $z = 1.52$. Output for comparing two proportions is given below. For "by hand" calculations, use the method described in Chapter 12. Alternatively, a chi-square test can be done and the relationship $z = \sqrt{\chi^2}$ used to determine the z-statistic. Output for the chi-square test is also shown.

Steps 3, 4, and 5: p-value = .064 (reported in output). Do not reject the null hypothesis. Based on these data it cannot be concluded that the proportions of Stars differ in the populations of Sheep and Goats.

```
                      Output for Exercise 15.49

   Test of two proportions:
   Sample      X      N  Sample p
   1          79    112  0.705357
   2          48     80  0.600000

   Estimate for p(1) - p(2):  0.105357
   95% lower bound for p(1) - p(2):  -0.00925990
   Test for p(1) - p(2) = 0 (vs > 0): Z = 1.52 P-Value = 0.064

   Chi-square test:
                Stars     Duds    Total
   Sheep          79       33      112
                74.08    37.92               Chi-Sq =   0.326 +  0.638 +
                                                        0.457 +  0.893 = 2.313
   Goats          48       32       80        DF = 1, P-Value = 0.128
                52.92    27.08

   Total         127       65      192
```

154

15.51 **a.** The conditions are met. In particular, only one of the expected counts for the twenty cells is less than 5. Expected counts are shown beneath observed counts in the output for part (b). Also, it is assumed the GSS sample represents a random sample from the population of U.S. adults.

b. Step 1: H_0: Religion and opinion about premarital sex are not related
  H_a: Religion and opinion about premarital sex are related

Step 2: $\chi^2 = 108.135$, df $= (5-1)(4-1) = 12$.

Steps 3,4, and 5: p-value ≈ 0. Reject the null hypothesis. Conclusion is that religion and opinion about premarital sex are related variables. Row percentages should be examined to determine how the religions differ.

```
                            Output for Exercise 15.51b

                              Almost
                Always        Always     Sometimes   Never       All

Protestant       221           54           98         288         661
                155.74        45.75       104.86      354.66      661.00

Catholic          45           17           54         179         295
                 69.51        20.42        46.80      158.28      295.00

Jewish             2            1            8          18          29
                  6.83         2.01         4.60       15.56       29.00

None              15           10           32         164         221
                 52.07        15.29        35.06      118.58      221.00

Other             20            7           12          41          80
                 18.85         5.54        12.69       42.92       80.00

All              303           89          204         690        1286
                303.00        89.00       204.00      690.00     1286.00

Cell Contents:        Count
                      Expected count

Pearson Chi-Square = 108.135, DF = 12, P-Value = 0.000

* NOTE * 2 cells with expected counts less than 5
```

15.53 **a.** H_0: $p_1 = .257$, $p_2 = .322$, $p_3 = .169$, $p_4 = .149$, $p_5 = .103$ (U.S. proportions)
  H_a: probabilities are not all as specified in the null hypothesis

b. Step 1: Hypotheses given in part (a)
  Step 2: $\chi^2 = 10.6$, df $= 5-1 = 4$

Expected counts, calculated as $2904 \times$ null p_i are:

Household Size	1	2	3	4	5
Expected	746.328	935.088	490.776	432.696	299.112

Steps 3,4, and 5: p-value $= .031$, from Excel CHIDIST(10.6), or with Table A.5, $.025 < p$-value $< .05$. Reject the null hypothesis. Conclude that the observed distribution of household sizes in the GSS survey is inconsistent with the U.S. distribution.

155

Note: This exercise illustrates the effect of a large sample size on a significance test. Notice that the sample proportions do not differ from the U.S. proportions by very much but due to the large sample size the difference achieves statistical significance.

15.55 $\chi^2 = \dfrac{N(AD-BC)^2}{R_1R_2C_1C_2} = \dfrac{194(80\times36-51\times27)^2}{131\times63\times107\times87} = 5.704$

15.57 **a.** H_0: Gender and perception of weight are not related
H_a: Gender and perception of weight are related
b. Reject the null hypothesis, because p-value = .000. There is a statistically significant relationship between gender and weight perception.
c. There are large contributions in the Underweight and Overweight categories for each sex. This reflects a large gender difference in those categories. Relevant condition percentages are: for females 30.2% (39/129) said overweight compared to only 3.6% of males (3/83), and for females only 2.3% (3/129) said underweight compared to 19.3% of males (16/83).

15.59 Between 0 and 5.99. In Table A.5, the value for df = 2 under column heading = .05 provides the answer. When the null hypothesis is true, 5% of the time the chi-square statistic will be $\geq$ 5.99. So, 95% of the time the statistic will be less than this value.

15.61 **a.** Too many cells have expected counts less than 5.
b. Step 1: H_0: Own eye color and eye color attracted to are not related
H_a: Own eye color and eye color attracted to are not related
Step 2: Test statistic is $\chi^2 = 15.5$, df = 4. Output is shown below, with expected counts beneath observed counts.
Steps 3, 4, and 5: p-value =.004. With Table A.5, .001< p-value <.005. Reject the null hypothesis. There is a statistically significant relationship between own eye color and eye color attracted to for the population of college students represented by this sample.

Output for Exercise 15.61b			
Rows: Own **Columns**: Attracted to			
	Brown	Blue	HazGr	All
Brown	30	22	19	71
	20.56	30.65	19.79	71.00
Blue	15	37	14	66
	19.11	28.49	18.39	66.00
HazGr	8	20	18	46
	13.32	19.86	12.82	46.00
All	53	79	51	183
	53.00	79.00	51.00	183.00
	--	--	--	--
Chi-Sq =	4.331 +	2.441 +	0.031 +	
	0.886 +	2.541 +	1.049 +	
	2.126 +	0.001 +	2.093 =	15.500
DF = 4, P-Value = 0.004			

c. The "contributions" to the chi-square statistic are shown in the calculation of the statistic at the bottom of the output. The largest two contributions are brown attracted to brown (4.331) and blue attracted to blue (2.541). In both cases, the observed count is much higher than the expected count. People appear to be attracted to people with their own eye colors more often than would be expected if the two variables were not related.

156

15.63 **a.** Homogeneity. The issue is whether the distribution of responses for satisfaction is the same for the two years.

H$_0$: Satisfaction with K-12 was the same for the populations of school parents in 1999 and 2000

H$_0$: Satisfaction with K-12 differed for the populations of school parents in 1999 and 2000

b. <u>Step 2</u>: Test statistic is $\chi^2 = 7.96$, df = 3.

<u>Steps 3,4, and 5</u>: *p*-value = .047. Reject the null hypothesis. Conclude that opinion differed in the two years. Examination of conditional percentages for the two years shows that a higher percentage were "completely satisfied" in 1999 than in 2000, while a higher percentage were "completely dissatisfied" in 2000 than in 1999.

Output is shown below, with expected counts beneath observed counts.

Output for Exercise 15.63			
Year =1999	2000	All	
Com Sat	125	87	212
	116.09	95.91	
Some Sat	155	133	288
	157.70	130.30	
Some Dis	41	34	75
	41.07	33.93	
Com Dis	7	17	24
	13.14	10.86	
All	328	271	599
Chi-Sq = 7.960	DF = 3,	P-Value = 0.047	

15.65 **a.** <u>Step 1</u>: H$_0$: gender and typical seat location are not related

H$_a$: gender and typical seat location are related

<u>Step 2</u>: $\chi^2 = 7.112$, df = 2

<u>Step 3</u>: *p*-value = .029 (given in Minitab output)

<u>Steps 4 and 5</u>: Reject the null hypothesis. Conclude that typical seat is related to gender. The conditional percentages by sex show that males are more likely than females to sit in the back and females are more likely than males to sit in the front.

Output for Exercise 15.65 (Observed counts and chi-square)			
Rows: Sex Columns: Seat			
	B	F	M	All
Female	22	38	93	153
Male	24	15	46	85
All	46	53	139	238
Chi-Square = 7.112, DF = 2, P-Value = 0.029			

Conditional percentages for seat location by sex are:

	Back	Front	Middle	All
Female	14.4%	24.8%	60.8%	100%
Male	28.2%	17.7%	54.1%	100%

b. A breakdown of the chi-square calculation is:

```
Chi-Sq =  1.939 +  0.453 +  0.149 +
          3.489 +  0.815 +  0.267 = 7.112
```

The contributions to chi-square are largest for the "Back" location for males (3.489) and females (1.939). This reflects a large difference between the males and females for the likelihood of sitting in the back.

15.67   Answers will vary.

15.69   **a**. The table of observed counts is the following:

Observed Counts for Exercise 15.69a

Rows: Livewhere   Columns: AlcMissClass

	No	Yes	All
Dorm	59	35	94
Off-campus	35	60	95
All	94	95	189

Cell Contents: Count

b. Row percents are the following:

Row percents for Exercise 15.69b

Rows: Livewhere   Columns: AlcMissClass

	No	Yes	All
Dorm	62.77	37.23	100.00
Off-campus	36.84	63.16	100.00
All	49.74	50.26	100.00

Cell Contents: % of Row

c. Students who lived off-campus were much more likely to have missed a class due to drinking alcohol. **Note**: In part, this difference is due to age as students who live off-campus are more likely to be older (and of legal drinking age.)

d. Null: Residence location and whether student has ever missed a class due to drinking alcohol are not related
 Alternative: Residence location and whether student has ever missed a class due to drinking alcohol are not related
e. p-value ≈ 0 (based on chi-square = 12.7 with df = 1). The result is statistically significant (at any level). We can infer that the two variables are related.

CHAPTER 16
ODD-NUMBERED SOLUTIONS

16.1   **a.** Appropriate. The response variable is quantitative and this is a comparison of independent groups.
.   **b.** Not appropriate. It's not a comparison of independent groups. There was only one group and all individuals listened to all five songs.

16.3   **a.** The null hypothesis was that the mean body mass is the same for the three age group populations. Using symbols, this would be written as $H_0: \mu_1 = \mu_2 = \mu_3$

        b. The *p*-value is small, below the usual .05 standard for significance, so it can be concluded that mean body mass is not the same for the three age group populations.

16.5   **a.** Mean height is not the same in the three seating location populations of female students
        b. Mean height increases from front to back.

16.7   **a.** Critical value (from table A.4) is 3.89. We can reject the null hypothesis because the F-statistic, 6.27, is greater than the critical value, 3.89.
        b. Cannot reject the null. (Critical value is 2.35.)

16.9   **a.** The population mean for the 60+ age group is different from the population means for the middle two age groups. (The confidence intervals for the differences do not include 0.) The population mean for the youngest age group is also different from the population means for the middle two age groups. We cannot conclude a difference between population means for the youngest and oldest age groups, and we cannot conclude a difference between the population means for the two middle age groups.
        b. There is 95% confidence that all six intervals capture the corresponding population parameters. And, there is a 100%–95% = 5% chance that at least one of the six intervals does not capture the corresponding population parameter. (In this situation, a parameter is a difference in population means.)

16.11   **a.** The data could be graphed using a comparative dotplot or a side-by-side boxplot (shown below). Both graphs show that measurements tend to be larger in labs 5 and 2 and tend to be smaller in lab 4. The data for labs 1 and 3 fall in between.

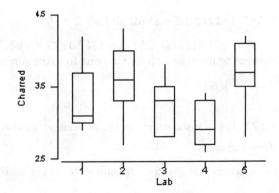

Figure for Exercise 16.11

        b. The necessary conditions are present. Variation is about the same for each lab, there are no extreme outliers and there is not extreme skewness. (It may appear from the boxplots that data for labs 1,3 and 5 are skewed, but remember that these plots are based on only 11 observations from each lab.)

159

16.13   **a.** H_0: $\mu_1 = \mu_2 = \mu_3$ versus H_a: not all μ_i are the same, where μ_i is the population mean testosterone level for all women in occupation group i.

   b. There are violations of the necessary conditions. There is an obvious outlier in the second group and two outliers in the first group as well. Also, there is greater variation within the advanced degree group than in the not employed group.

16.15   The completed table is:

Source	DF	SS	MS	F
Between groups	5	40	8	1.333
Error	10	60	6	
Total	15	100		

The general relationships used are: MS = SS/df, F = MS Between/MS Error, and the total df and total SS are the totals of the corresponding quantities for Between and Error.

16.17   The sample means would all equal the same value.

16.19   **a.** The completed table is:

Source	DF	SS	MS	F	P
Caffeine	2	61.40	30.70	6.18	0.006
Error	27	134.10	4.97		
Total	29	195.50			

In the Caffeine row, df = k–1=3–1=2 and $MS = SS$/df = 61.40 /2 = 30.70

In the Error row, df = N–k = 30–3 =27 and $MS = SS$ / df = 134.10 / 27 = 4.97. Equivalently, the df and the SS can be determined by subtracting the df and SS for Caffeine from the corresponding values for Total.

Finally, $F = \dfrac{MS\ Caffeine}{MSE} = \dfrac{30.70}{4.97} = 6.18$

b. $s_p = \sqrt{MSE} = \sqrt{4.97} = 2.23$. This statistic estimates the population standard deviation of the response variable (for any of the caffeine amounts).

c. The p-value is small so we can conclude that the population means are not the same for the three caffeine amounts.

16.21   Each interval has the form $\bar{x}_i \pm t^* \dfrac{s_p}{\sqrt{n_i}}$ which in this case is $\bar{x}_i \pm 2.01 \dfrac{0.4058}{\sqrt{11}}$, or $\bar{x}_i \pm 0.246$. Use Table A.2 to find the multiplier. The degrees of freedom are the Error df = 50.

The five intervals are:

lab 1, 3.3364 ± 0.246; lab 2, 3.6 ± 0.246; lab 3, 3.3 ± 0.246; lab 4, 3.0 ± 0.246; lab 5, 3.6455 ± 0.246

16.23 **a.** $\bar{x} = 8$, $\bar{x}_1 = 4$, $\bar{x}_2 = 11$, $\bar{x}_3 = 9$.

b. $SS\ Groups = \sum n_i (\bar{x}_i - \bar{x})^2 = 3(4-8)^2 + 3(11-8)^2 + 3(9-8)^2 = 48 + 27 + 3 = 78$

c. $SS\ Total = \sum (x_{ij} - \bar{x})^2 =$

$(6-8)^2 + (4-8)^2 + (2-8)^2 + (10-8)^2 + (14-8)^2 + (9-8)^2 + (9-8)^2 + (12-8)^2 + (6-8)^2 = 118$

d. $SSE = SS\ Total - SS\ Groups = 118 - 78 = 40$.

e. $F = \dfrac{\dfrac{SS\ Groups}{k-1}}{\dfrac{SSE}{N-k}} = \dfrac{\dfrac{78}{3-1}}{\dfrac{40}{9-3}} = 5.85$, $df = (k-1,\ N-k) = (2, 6)$.

16.25 **a.** The completed table is:

Source	DF	SS	MS	F	P
Treatment	3	150	50	5	0.01
Error	20	200	10		
Total	23	350			

16.27 **a.** H_0: median ratings are the same for the four populations of hometown types
   H_a: population medians are not all the same

16.29 **a.** Ordinal because an arbitrary rating scale was used.
b. The null hypothesis is that the median response is the same for the populations of college men and women represented by this sample. The alternative hypothesis is that the population medians are not equal.
c. The overall median was 12 (given in the next to last line of output). Among women, 104 gave an answer less than or equal to the overall median while 46 did not, so the proportion is 104/(104+46) or 69.3%. Among men, 37 gave an answer less than or equal to the overall median while 48 did not, so the proportion is 37/(37+48) or 43.5%.
d. The null hypothesis defined in part (b) can be rejected. In other words, a statistically significant difference has been observed. The p-value for the test is given as 0.000 in the first line of output. Regarding responses to this question, it would be reasonable to conclude that there is a difference between females and males in the population of college students represented by this sample.   Notice, by the way, that the sample evidence is that females tend to respond closer to the "personality most important" end of the scale.

16.31 **a.** H_0: median testosterone levels are the same for the three populations of occupational groups
   H_a: population medians are not all the same
b. Decide in favor of the alternative hypothesis (p-value=0.002). The population median testosterone level differs for at least one of the occupational groups.
c. The sample medians are 3.4 for the "advanced degree" group, 2.2 for the "no advanced degree" group, and 2.0 for the "not employed group." Along with the figure for Exercise 16.6, these sample medians indicate that the principal difference is that measurements tend to be higher in the "advanced degree" group than in the other two groups. There may not be much difference between the "no advanced degree" and "not employed" groups.

16.33 **a.** Gender and Greek membership (yes or no) are interacting variables. The size of the mean difference in mean hours spent studying for Greeks versus non-Greeks depends upon gender.

16.35 **a.** Males (described by the upper line in the graph).
b. For both males and females, the mean rating is clearly lower for "Big city" than for the other three types of hometown.

c. There probably is not a significant interaction (either statistically or in practical terms). The difference between males and females is about the same for each hometown type. Equivalently, the pattern of differences among the four hometown types is about the same for each sex.

16.37 **a.** The amount of difference between mean GPA of men and women would depend upon seat location.
b. The amount of difference between men and women would be the same in each seat location.

16.39 **a.**

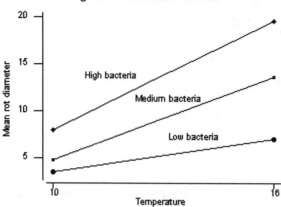

Figure for Exercise 16.39a

b. As the bacteria amount increases, the difference in mean rot for 10°C and 16°C increases as well. Notice in the plot that the line for high bacteria has a steeper slope than the line for low bacteria.

16.41 The confidence intervals have unequal widths because the sample sizes differ for each group. The pooled estimate of the standard deviation is used for each group, but the standard error of the mean is found by dividing that standard deviation by the square root of the sample size. Smaller sample sizes produce wider confidence intervals.

16.43 **a.** The null hypothesis is that the population mean violent behavior scores for students represented by the sample are the same for the five television watching groups. The alternative hypothesis is that the mean violent behavior scores are not the same for all five groups. These hypotheses are written as

H_0: $\mu_1 = \mu_2 = \mu_3 = \mu_4 = \mu_5$

H_a: not all μ_i are the same.

b. The intended population is all school children of this age in the U.S although this could be questioned because the sample includes children only from Ohio.
c. Decide in favor of H_a. Not all the means are the same.
d. The cause and effect might go in the other direction. Perhaps children prone to more violent behaviors also like to watch more television. This was an observational study, and there are many other possibilities for confounding variables, such as lack of parental attention.

16.45 There may be an interaction. The difference between boys and girls is largest in the <1 hour of television group and smallest in the 6+ hours of television group.

16.47 Regression methods could be used. Both variables are quantitative. An advantage would be that regression provides an estimate of the relationship between caffeine amounts and mean tapping speed.

16.49 $SSE = \sum (n_i - 1)^2 s_i^2 = (162 - 1)(7.15)^2 + (67 - 1)(6.43)^2 + (105 - 1)(6.45)^2 \approx 15,286$

162

$$MSE = \frac{SSE}{N-k} = \frac{15286}{334-3} = 46.18$$

$$s_p = \sqrt{MSE} = \sqrt{46.18} = 6.796$$

16.51   **a.** 0.0846 to 0.4824

b. 0.1164 to 0.4506

c. There are 3 pairs. Use $1-(.05/3) = 0.9833..$ as the confidence level for each interval.

d. Fisher gives narrowest interval and Bonferroni gives the widest. Fisher gives 95% confidence for each interval, whereas Bonferroni and Tukey give 95% family confidence (so higher confidence for each individual interval).

16.53   **a.** The aroma ratings are generally higher for wines from region 3 than for regions 1 and 2.

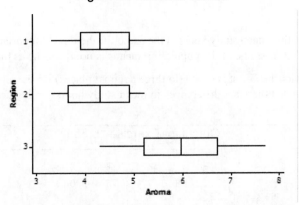

Figure for Exercise 16.53a

b.

Region	N	Mean	Std. dev.
1	17	4.359	0.685
2	9	4.278	0.676
3	12	5.967	0.962

c. Null: Population mean aroma ratings are the same for the three regions
Alternative: At least one of the population mean aroma ratings is not the same as the others.

The *p*-value for the *F*-test is 0.000 (to three decimal places), so we reject the null hypothesis.

We conclude that there are differences among the population mean aroma ratings for the three regions.

```
                  Output for Exercise 16.53c

    Source   DF      SS      MS      F      P
    Region    2  22.011  11.006  18.05  0.000
    Error    35  21.343   0.610
    Total    37  43.355
```

163

16.55 **a.** Students who say that religion is very important in their lives tend to study more. The median and the third quartile are greater for this group than for the other two groups.

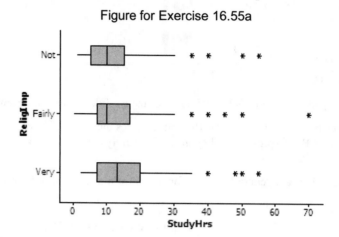

Figure for Exercise 16.55a

b. Null: Population mean study hours is the same for the three "importance of religion" groups. Alternative: At least one of the population means is not the same as the others.

The p-value for the F-test is 0.000 (to three decimal places), so we reject the null hypothesis. We conclude that there are differences in mean study hours for the three "importance of religion" groups

Output for Exercise 16.55b					

```
Source      DF       SS      MS     F       P
ReligImp     2    1721.4   860.7  9.77   0.000
Error      683   60183.6    88.1
Total      685   61905.1
```

c. The standard deviations for the three groups are not notably different (they're 8.5, 9.0, and 11.3. There is some skewness in the data, and a few outliers, but with large sample sizes these factors are not likely to cause serious difficulties.

16.57 **a.** The means and medians decrease as the years of education increases.

	n	Mean	Median
Not HS	134	3.970	3
High school	506	3.156	3
Junior college	55	2.436	2
Bachelor's	140	2.200	2
Graduate	70	1.843	1

b. Null: Population median television watching amount is the same for the five degree groups.
Alternative: At least one of the population medians is not the same as the others.

For both the Kruskal-Wallis test and the Mood's median test, the p-value = 0.000 (to three decimal places). We can conclude that there are differences among the five educational degree groups with respect to median self-reported television watching amounts.

c. Null: Population mean television watching amount is the same for the five degree groups.
Alternative: At least one of the population means is not the same as the others.

For the F-test, $F = 16.22$ and the p-value $= 0.000$ (to three decimal places). We can conclude that there are differences among the five educational degree groups with respect to mean self-reported television watching amounts.

16.59 **a.** There may possibly be a weak interaction. The difference between males and females is greater for students who sit in the back, compared to students who it in other locations.

Figure for Exercise 16.59

b. In general, males miss more classes than females. Mean numbers of classes missed increase as we move toward the back of the classroom.

CHAPTER 17
ODD-NUMBERED SOLUTIONS

17.1   **a.** No, the entire population already has been measured.

b. Yes, if we assume that the mechanism causing geyser eruptions stays the same over time it would be reasonable to use data from these two weeks to represent the larger population of all eruptions (ever) of the Old Faithful geyser.

Note: The pattern seems to have changed in recent years. According to the National Park Service website http://www.nps.gov/yell/planyourvisit/oldfaithfaq.htm, "In the past, Old Faithful displayed two eruptive modes: short duration eruptions followed by a short interval, and a long duration eruption followed by a long interval. However, after a local earthquake in 1998, Old Faithful's eruptions are more often of the long duration, long interval type. If an eruption lasts less than 2.5 minutes than there will be a 60 minute interval. If an eruption lasts more than 2.5 minutes, there will be a 90 minute interval."

17.3   No. This is an observational study. Children cannot be randomly assigned to adoptive parents.

17.5   The study needs to be a randomized experiment, in which treatments are randomly assigned to participants.

17.7   **a.** The null hypothesis is that it will not rain and the alternative is that it will rain. Type 1 error is that it does not rain, but you take an umbrella. Type 2 error is that it does rain but you do not have an umbrella. The likelihood of rain was provided by your weather website. Here is the table:

	Likelihood	H₀ Chosen (no umbrella)	Hₐ Chosen (Umbrella)
H₀ True (No rain)	60%	No loss	Carry umbrella that isn't needed.
Hₐ True (Rain)	40%	You get wet.	Umbrella keeps you dry.

b. Reasoning is what counts here, but Type 2 error is probably more serious because you would get wet. So you probably would want to act as if the alternative hypothesis will be true and carry an umbrella.

17.9   **a.** Yes, because the study was a randomized experiment.
b. Yes, those men are probably representative of healthy men in the same age group, and possibly a larger population.

17.11   Answers will vary, but make sure the study is an observational study rather than a study in which random assignments are used.

17.13   Answers will differ for each student, but make sure that the study either used random assignment of the treatments to participants, or randomized the order of treatments if each individual received all treatments.

17.15   **a.** Yes, the wording is justified. The quote only says that the "tobacco-control program is associated with a reduction in deaths." While there is a subtle (or perhaps not so subtle) implication that the tobacco-control program caused the reduction in deaths, the word "association" means only that the researchers observed that deaths went down after the tobacco-control program was instituted. A cause-and-effect conclusion is not justified because the reduction in deaths might have been the result of changes in other health-related factors that occurred after the program was begun.

b. Answers will differ for each student. An example of a headline that accurately describes the result is "Heart disease deaths down since start of anti-smoking measures." An example of an unjustified headline is "Anti-smoking measures cause drop in heart disease deaths."

17.17   **a.** Explanatory variable: Calcium supplement or placebo. Response variable: Blood pressure.
b. Yes, the words "calcium lowers blood pressure" imply that calcium *causes* lower blood pressure.

17.19   **a.** Explanatory variable is daily tea consumption and the response variable is conception (conceived or not).

167

b. Yes, the headline implies that drinking tea is responsible for doubling the chances of conception.

17.21   Study #2 apparently was an observational study. The article does not state that the investigators assigned walking amounts to the men, so we can presume that the investigators only observed these amounts and that each man was free to choose his own amount of walking. Because this is an observational study, the headline is not justified (see Rule for Concluding Cause and Effect on page 652). There may be confounding variables that contribute to the lower death rate for those who walk more than 2 miles per day. For instance, they may have healthier diets than men who walk lesser amounts. Also, the cause and effect might, to some extent, go in the opposite direction. Healthier men would be able to walk more.

17.23   Study #3 apparently was an observational study. The article says that that the researchers asked the women to "record information concerning their daily dietary intake" so the investigators only observed tea consumption. The researchers did not assign amounts of tea drinking to the women. Because this is an observational study, the headline is not justified (see Rule for Concluding Cause and Effect on page 652). There may be confounding variables that contribute to the higher odds of conception for those who drank more tea.

17.25   Based on the information given, the results can probably be applied to the larger population of African-American teens in the United States. While the sample was not a random sample from this population, it seems reasonable to assume that the sample would be representative of the larger population for this issue. It might also be argued that for the question of interest, the results might apply to all teens (not only African-Americans).

Note: In a part of the article not given in the text, it is reported that the teens in the study all had high blood pressure prior to the study so the population of interest actually is the population of African-American teens who have high blood pressure.

17.27   Yes, the results probably can be applied to a larger population. We are not told who was in the sample, but there is no indication that it was an unrepresentative group with regard to diet (and tea drinking) and conception. The Fundamental Rule for Using Data for Inference probably holds for the question of interest here.

17.29   **a.** "Significance" most likely refers to statistical significance. The article described a research paper published in the American Journal of Public Health, and in that type of context the word "significance" usually has to do with the results of statistical significance tests. The null hypothesis is that energy and fat intake are not related to tea consumption. A one-sided alternative hypothesis consistent with the research conclusion is that energy and fat intake increase as tea consumption increases. A two-sided alternative hypothesis is that energy and fat intake are related to tea consumption.

b. Energy and fat intake are confounding variables. They are *related* to the explanatory variable (tea consumption) and may possibly *affect* the response variable (odds of conception). See the definition of confounding variables on page 191.

17.31   **a.** An association was observed in the sample, but it was not strong enough to be statistically significant for the sample size used.

b. Two possibilities are suggested in item 10 in Section 17.7, and both possibilities exist here. First, the conclusion of the present study might be incorrect because the sample size was too small to be conclusive. Second, the conclusion of the previous research that concluded significance might be wrong. We aren't given enough information about the details of the prior research to be able to judge this. One question of interest is whether only one previous study found a relationship or whether several previous studies found a relationship. If several previous studies found a relationship, rather than only one, it is likely that the error is in the present study.

17.33   **a.** Amount walked per day (less than one mile, between one and two miles, more than two miles) and whether the man died during the study (yes, no).

b. A test for the difference in two proportions would be appropriate. Define p_1 to be the proportion of men who would die during a time period similar to this study, for the population of men like the ones in the study who walk less than one mile per day. Define p_2 to be the equivalent proportion for men who walk more than two miles per day. The hypotheses would then be $H_0: p_1 - p_2 = 0$ and $H_a: p_1 - p_2 < 0$.

168